Structural Engineering: A Very Short Introduction

VERY SHORT INTRODUCTIONS are for anyone wanting a stimulating and accessible way in to a new subject. They are written by experts, and have been translated into more than 40 different languages.

The Series began in 1995, and now covers a wide variety of topics in every discipline. The VSI library now contains over 350 volumes—a Very Short Introduction to everything from Psychology and Philosophy of Science to American History and Relativity—and continues to grow in every subject area.

Very Short Introductions available now:

Available soon:

For more information visit our website

www.oup.com/vsi/

David Blockley

STRUCTURAL ENGINEERING

A Very Short Introduction

OXFORD
UNIVERSITY PRESS

OXFORD

UNIVERSITY PRESS

Great Clarendon Street, Oxford, OX2 6DP,
United Kingdom

Oxford University Press is a department of the University of Oxford.
It furthers the University's objective of excellence in research, scholarship,
and education by publishing worldwide. Oxford is a registered trade mark of
Oxford University Press in the UK and in certain other countries

First edition published in 2014
Impression: 6

Published in the United States of America by Oxford University Press
198 Madison Avenue, New York, NY 10016, United States of America

British Library Cataloguing in Publication Data
Data available

Library of Congress Control Number: 2014938488

ISBN 978-0-19-967193-9

Printed in Great Britain by
Ashford Colour Press Ltd, Gosport, Hampshire

Contents

Preface

If you have ever wondered how a skyscraper, like the Shard in London, a big bridge, like the Golden Gate in San Francisco, a jumbo jet like the A380, or a cruise liner, like the *Queen Elizabeth*, stand tall: if you have ever wondered where the architecture stops and the engineering begins—then this is a book for you. We will explore in non-technical language how large man-made structures are created.

The purpose of the book is three-fold. First, I aim to help the general reader appreciate the nature of structure, the role of the structural engineer in man-made structures, and understand better the relationship between architecture and engineering. Second, I provide an overview of how structures work: how they stand up to the various demands made of them. Third, I give students and prospective students in engineering, architecture, and science access to perspectives and qualitative understanding of advanced modern structures—going well beyond the simple statics of most introductory texts.

Inevitably in such a short volume choices have to be made about what not to include. There will be gaps—in particular I have made no attempt to describe the everyday life and work of a structural engineer, the science of materials is limited, and I make only brief references as to how structures are actually put together. I have included some recommendations for further reading.

Everything has structure. The function of structure is to provide the form and shape on which other functions can operate. Natural structures vary from the very smallest part of an atom to the entire cosmology of the universe. Man-made structures include buildings, bridges, dams, ships, aeroplanes, rockets, trains, cars, and fairground rides, and all forms of artefacts—even large sculptures like the Angel of the North in the UK.

Structure is the difference between a random pile of components and a fully functional object. Through structure the parts connect to make the whole.

Engineering is the 'turning of ideas into reality'. Structural engineering is a critical part of the fulfilling of some of our most basic human needs—particularly for shelter, protection, and travel. It has evolved from the common-sense building of primitive huts, bridges, and weapons of ancient history, through the craft skills of medieval master masons who built castles, cathedrals, and country houses, to the latest and most sophisticated use of the science of structures and materials.

The wide range of different industries in which structural engineers work includes construction, transport, manufacturing, and aerospace. Each industry has its own particular ways of doing things. The book attempts to get behind those differences to the essential common core.

The work in a typical structural engineering design consultancy might be 30 per cent on design discussions; 30 per cent on structural analysis; 30 per cent on determining construction details; and 10 per cent on specifying, supervising, and checking the work. The structure of the book does not follow that pattern—instead we begin in Chapter 1 to trace the close and often controversial relationship between architectural and structural form and function. This clarification is important because so many people, including much of the media, see structure as architecture. We will see that architects rely on engineers to make their ideas work, in other words, to stand up safely and operate successfully. However, just as an architectural form may not function well structurally so an efficient structural form may not be good architecturally. The best structures are a harmony of architecture and engineering—where form and function are one and the flow of forces is logical. We also learn in Chapter 1 how you can develop your own understanding of structural form with some very simple experiments you can do at home. In Chapter 2 we examine the three requirements for good structure set by the Roman Vitruvius long ago: resilience, purpose, and delight. We see that force pathways are degrees of freedom and that so-called form-finding structures are exciting and innovative examples of the fusion of engineering and architecture. We find that structures are naturally lazy because they contain minimum potential energy. Chapter 3 gives us the historical perspective necessary to understand the modern controversies as we trace the story of the emergence of the master mason and the increasingly separate roles of the architect and the engineer. We see how, from the Renaissance onwards, specialisms grew through new scientific knowledge, new materials, and new demands for structures. We begin to understand how the engineering profession developed into a science based capability to create some very large and breathtaking structures. Chapter 4 then builds on that background to help us understand better the science of how structures work. By that I mean how they resist all of the demands made on them by forces of self-weight, people moving about, wind that may blow a hurricane and ground that may vibrate in an earthquake. These forces have to 'flow through' the components of the structure rather as water flows through a set of pipes. We introduce some of the major theoretical principles that are the basis for modern computer analysis methods. In Chapter 5 we examine the commonalities and differences between four types of very large impressive structures—a jumbo jet, a cruise ship, a long span bridge, and a skyscraper building. Finally in Chapter 6 we ask questions about risk because structures are safety critical—when they fail people may be killed. We recognize that no human activity is risk free. We ask how safe is safe enough. We identify a relatively new philosophy called 'systems-thinking' to examine how we might provide structures that will be resilient to the extreme weather events caused by changes in our climate should the worst scientific predictions come to pass.

Structural engineering is an important part of almost all undergraduate courses in engineering. This book is novel in the use of 'thought-experiments' as a straightforward way of explaining some of the important concepts that students often find the most difficult. These include virtual work, strain energy, and maximum and minimum energy principles, all of which are basic to modern computational techniques. The focus is on gaining understanding without the distraction of mathematical detail. The book is therefore particularly relevant for students of civil, mechanical, aeronautical, and aerospace engineering but, of course, it does not cover all of the theoretical detail necessary for completing such courses.

Structural engineering is a dominantly male profession. An increasing number of women are being recruited and rising to senior levels. Structural engineers, their professional bodies, the universities, and governments around the world are actively campaigning to recruit more women (see for example <http://www.wisecampaign.org.uk/> and <http://raeng.org.uk/news/publications/list/reports/Inspiring_Women_Engineers.pdf>). In this spirit I have used the feminine gender when needed.

I would like to thank a number of people who have helped me enormously. Anne Thorpe and Oksana Kasyutich have both read the whole book and made several suggestions to help make it more accessible to the lay reader. My own specialism within structural engineering is in construction and so the invaluable help of a number of structural engineers in naval architecture and aerospace engineering has been crucial. In particular I am indebted to Emeritus Professor John Caldwell of the Department of Naval Architecture at the University of Newcastle for his close reading of my references to all things naval and for making many useful corrections and suggestions. Thanks to Vaughan Pomeroy, ex-Technical Director of Lloyd's Register and now Visiting Professor at the University of Southampton, who has read the whole book and made a number of corrections and suggestions. I am grateful to Danny Heaton, retired from Airbus, also for reading the entire book and particularly for helping me understand aircraft regulations. I thank Professor Paul Weaver of the Department of Aerospace at the University of Bristol for spotting at least one error and for checking out my references to aircraft structures. I am grateful for the close reading by Emeritus Professor Mike Barnes of the Department of Civil Engineering at the University of Bath and for making many pertinent and useful comments for improvements. I thank Michael Dickson for some useful advice and Simon Pitchers for valuable conversations about the relationships between structural engineering and architecture. Thanks go to David Elms for pointing me to the mathematical proof that you can find a stable position for a four-legged table on a rough floor and for consequent discussions that led to the explanation in the Reference Section to Chapter 4. I thank Robert Gregory for his reading of the text and permitting me to include his photograph of a spider's web.

Finally thanks are due to Latha Menon at Oxford University Press who supported the idea for the book, Jenny Nugee, Emma Ma, Carol Carnegie, Subramaniam Vengatakrishnan, Joy Mellor, and Kay Clement.

Last by no means least, thanks to my wife Karen for her unfailing love and support, and endless cups of tea.

List of illustrations

Structural Engineering

Chapter 1
Everything has structure

Your bones and muscles give you structure—form and shape. Structural engineering is about the bones and muscle of man-made things like buildings, bridges, ships, and aeroplanes. Structural engineers make buildings and bridges stand up safely. The purpose of this book is to help you to look 'with new eyes' at the man-made structures around you and hence to better understand why they are as they are.

In this first chapter we will look at the fundamental nature and role of structure and the relationship of structural engineering with other engineering disciplines and with architecture. We will see that decision making is driven by the purpose of a man-made structure and how we realize 'fitness for purpose'. We will examine the need to understand how forces flow through a structure in order to ensure it meets its primary purpose of being strong and safe whilst at the same time meeting many other needs such as affordability and aesthetic, regulatory, and environmental criteria. Lastly we will investigate some simple experiments that help us to understand how structures work.

Your musculoskeletal system is not just your bones and muscles—it includes all of your connecting tissue such as the joints, ligaments, and tendons which help you move around. On it are hung all of your other bits and pieces, such as your heart, brain, liver, etc. Without structure you would just be a blob of jelly—structure supports who you are and how you function. Likewise the structure of a building isn't just beams and columns—it includes all of the connecting material such as the joints, slabs, welds, and bolts which keep it all together. On it are hung all of the other parts of the building such as the equipment for heating, lighting, communication, and all of the furniture, fixtures, and fittings. Without structure a building would just be a random pile of components—the function of structure is to support all the other functions of the building.

Everything in the natural world has structure, from the very small, like the carbon 60 molecule, through plants and insects (Figure 1) to the very large, such as mountains and indeed the whole Universe. The carbon atom is ubiquitous to life and at the heart of new

1. **Spider's web**

engineered materials. Insects and animals build structures—nests, burrows, and dens provide protection and places to rear the young. The spider in Figure 1 has woven a remarkable web to catch its prey—the silk has a tensile strength greater than the same weight of steel and with much greater elasticity. Beavers are another of nature's builders—they make dams for protection and to trap food.

Structure is the connecting of parts to make a whole—and it occurs at many different levels. Atoms have structure. Structures of atoms make molecules, structures of molecules make tissue and materials, structures of materials make organs and equipment, and so on up a hierarchy of different levels, as shown in Figure 2.

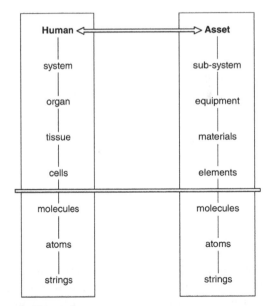

2. Structural hierarchy

Within this hierarchy of structure, man-made objects vary from the very small, like a nanotube or a silicon chip, to the very large, like a cruise ship or a jumbo jet. Whereas natural structures have evolved over aeons, man-made structures have to be imagined, designed, and built through our own efforts. As we shall see in Chapter 3 this purposeful activity of structural engineering has a long human history which is closely interconnected with the evolution of our understanding of science.

The intimacy of engineering and architecture

Structural engineering is seen by many as a specialty within civil engineering—concerning the built environment of buildings and bridges—but actually its scope is much wider. Mechanical and aeronautical engineers, and indeed all other engineers regardless of specialism, require a knowledge of structure. Structural engineering is part of the design, construction, repair, conversion, and conservation of sports stadia, reservoirs, dams, off-shore platforms, and fairground rides. Structural engineering knowledge and skills are essential for the design of machinery, cranes, trains, cars, buses, boilers and pressure vessels, medical equipment, ships, oil rigs, aeroplanes, rockets and space satellites, and electronic equipment. They are needed for deployable structures such as solar arrays and antennas on spacecraft which are stowed compactly for launch and then unfolded in space. Structural engineering is also important for less prosaic forms of man-made artefacts such as large sculptures like the Angel of the North.

The story of structural engineering is also closely bound up with that of architecture, and the two often are confused because both are concerned with form or shape—but from different points of view. We will distinguish them by referring to structural and architectural form. In brief, if you were a building, then the architect would decide your architectural form—your gender, what you look like, whether you are blonde or brunette, your body

4

shape and appearance. However, the architect would not decide what is necessary to make the different parts of your body function as they should—that is the job of various kinds of engineer. Structural engineers must ensure that the structural form is sufficient to carry the forces imposed on it safely—we will begin to examine that idea in the next section and then develop it more deeply in subsequent chapters. All roles require the skills of synthesis (designing), analysis (understanding), and practice (turning ideas into a reality)—but in different mixes.

In the widest sense engineering is conceiving, designing, constructing, operating, and eventually decommissioning something to fulfil a human need. We may talk about how to engineer a solution to a problem—we may engineer an agreement, a deal, or a plan or even a 'bright future' for ourselves. Architecture, on the other hand, is about devising, designing, planning, and supervising the making of something. We may speak of the architecture of the internet to refer to the way the highly interconnected network of computers fits together—we may choose architectural plants to provide structure in a garden.

In a narrower sense dictionaries commonly define architecture as the art and science of designing and erecting buildings and other physical structures. Architectural form concerns the sense and use of space, occupancy by people, symbolism and relationship to setting: it can be decorative and sculptural. Architects who design buildings are not engineers and rarely have the level of scientific knowledge required of professionally qualified engineers. The role of an architect is to understand and fulfil the needs of a client for the ways in which a building is to be used and how it will look—in other words its overall form, appearance, and aesthetic effect.

Another reason for the confusion between engineering and architecture is that some engineers call themselves architects. For example the engineers who design and build ships call themselves naval architects—there is no distinct non-engineering profession of architecture as there is in construction. Marine engineers design and operate on-board machinery.

Historically, in the aerospace and defence industries the term architect has not been used. However, in the latter part of the 20th century the engineers who design, build, and maintain the avionic systems that control an aeroplane, and other engineers involved with information technology and complex logistics have decided to call themselves 'systems architects' in order to address what they see as the shortcomings of traditional methods of engineering.

As we will discover in Chapter 3 the dual roles of the architect and engineer developed out of the single role of medieval master mason. The inevitable historical separation of the two professions has been unfortunate, divisive, and often dysfunctional. For example it has resulted in the over-simplified contrasting impression held by many that architecture is a form of creative art while engineering is 'merely' a craft, trade, or occupation that relies on manual skills.

Good architecture and good engineering are both arts requiring science—but they are aimed at different purposes. Art is difficult to define but is a power of the practical intellect, the ability to make something of more than ordinary significance. Science is a branch of knowledge which is systematic, testable, and objective—science is what we know. When architecture and engineering get artificially separated the outcomes may not be as they should be. For example a company or a developer investing in a new building such as an office block might appoint an architect to develop a scheme proposal to meet the needs of the client. If that is done, as has been all too often the case, without the involvement of properly qualified engineers then, later, when the project gets underway, there will inevitably be practical problems. In the worst building projects architects specify structural forms that may simply be unbuildable or unnecessarily expensive to build. It follows that in the best and most successful building projects architects and engineers work together right from the start. Good structural design can leverage orders of magnitude saving in the cost of construction.

Three perspectives

Let's think about what is required of a structure from three different basic perspectives—purpose, material, and form. Purpose is about *why* we need it and *how* we decide if it is 'fit for purpose'. Material is about *what* it should be made of. Form is about *what* shape it should be.

The various purposes and requirements of a structure are the first focus of the design team. Only then can they decide how the structure should function so that it is fit for those purposes and meets those requirements. Clearly bridges must carry traffic, buildings must enable people to live and work, and ships and aeroplanes must transport people and cargo. What will be the demands on the structure and how will it resist all of the forces of self-weight, people moving about, wind that may blow a hurricane, and ground that may vibrate in an earthquake? The team must examine how these forces will 'flow through' the components of the structure—rather as water flows through a set of pipes. In every case the requirements are embedded in aesthetic, technical, business, regulatory, environmental, social, cultural, and historical contexts that constrain possible solutions. A purpose and function defined without recognizing all of these constraints will be partial.

For example a strong but ugly ship is inadequate, although aesthetics is rarely a consideration for dredgers, bulk carriers, and workboats. A weak but beautiful ship is worse, because strength is a necessary requirement. A ship that is fitted out in a way inappropriate for its designated use will not be successful. For example a cruise liner has to have attractive internal architecture

7

with swimming pools on the upper decks—and, as we will see later, that is not the best location for overall buoyancy. So it is important to remember that whilst structural strength is not sufficient, it is necessary. Beyond that, the structure must be such that it will deliver on the need and be 'fit for purpose' in all aspects. A high quality aeroplane is one that fulfils its purpose, from all angles and points of view.

All structures span a gap or a space of some kind and their primary role is to transmit the imposed forces safely. A bridge spans an obstruction like a road or a river. The roof truss of a house spans the rooms of the house. The fuselage of a jumbo jet spans between wheels of its undercarriage on the tarmac of an airport terminal and the self-weight, lift and drag forces in flight. The hull of a ship spans between the variable buoyancy forces caused by the waves of the sea.

To be fit for purpose every structure has to cope with specific local conditions and perform inside acceptable boundaries of behaviour—which engineers call 'limit states'. For example houses in Switzerland and Canada need a very strong roof structure to deal with heavy snowfall, whilst houses in the Caribbean need roof structures to cope with hurricane winds. Bridges need to carry different kinds of loads such as pedestrians, cars, trucks, and high speed trains. A cruise ship taking thousands of passengers and crew into the rough seas and cold weather of the Arctic Circle has to operate safely. However, no one thing can be absolutely safe and passengers are aware that sea conditions can create safety concerns. There is inevitably an element of risk in any human activity since we simply cannot know everything about everything. It follows that one of the main challenges facing structural engineers is to manage the level of risk to the integrity of the structure of the ship and hence therefore to the lives of all the people on board. Of course a ship can run into problems for many different reasons varying from hitting rocks due to a navigational error to losing mechanical power—so the safety of the structure of the ship is necessary but not sufficient for the safety of the ship as a whole. In other words a ship structure must be safe but overall safety of the ship depends on many other factors as well.

Safety is paramount in two ways. First, the risk of a structure totally losing its structural integrity must be very low—for example a building must not collapse or a ship break up. This maximum level of performance is called an ultimate limit state. If a structure should reach that state for whatever reason then the structural engineer tries to ensure that the collapse or break up is not sudden—that there is some degree of warning—but this is not always possible and sudden collapses do occur. Second, structures must be able to do what they were built for—this is called a serviceability or performance limit state. So for example a skyscraper building must not sway so much that it causes discomfort to the occupants, even if the risk of total collapse is still very small. Likewise bridges mustn't wobble so much that people can't walk easily, even if the bridge is very unlikely to fall down. These two limit states, the ultimate and serviceability, are boundaries—just like the boundaries you have when you are in what some experts call your 'comfort zone'. Go beyond those boundaries and out of your comfort zone and you feel uncomfortable—but you can cope. You have exceeded a serviceability limit state.

It is perhaps understandable why engineers have, in the past, seen nature as a kind of cunning adversary. There are a myriad possible ways in which natural forces could 'attack' a structure and push it through a limit state boundary—imagining them all requires a certain kind of creative thinking. If a structure has a weak spot then eventually nature will find it. For example, if you own a house then you probably appreciate that if there is a way in which rain can get in—it will. Eventually you will discover a damp patch. If you find a problem in your timber roof structure, such as a sizeable infestation of woodworm, then unless you get it treated the roof will eventually be so seriously weakened that it could

9

collapse. Ships are the largest man-made mobile structures, and the essential task of structural engineers and naval architects is to ensure that the risk of any weakness occurring during the working life of a ship is kept to an acceptably low level. The equivalents of woodworm in man-made materials are various forms of deterioration such as rust and corrosion, cracks in brittle materials such as glass and brickwork. Indeed, even ductile steel can crack in a brittle manner due to a phenomenon called metal fatigue (Chapter 5). Just as you must maintain the structure of your house, so engineers have to be vigilant in spotting and dealing with threats to ships or aeroplanes.

This battle to control nature was perhaps the driving force behind one of the first definitions of civil engineering. In 1828, Thomas Tredgold defined it as '... the art of directing the great forces of power in nature for *the use and convenience* of man ...' (my italics). Now we see this definition as completely inappropriate—nature is not an adversary to be tamed for our use and convenience. Most engineers recognize that we humans are part of the natural world and not separate from it. We realize that our interests and those of the natural world are closely interlinked. We should not strive to dominate nature but rather to work in partnership with it.

Another significant risk to safety is that people do make errors—that is just part of being human—so all work must be appropriately supervised, checked, and corrected where necessary. Errors may vary from simple slips or lapses (like an arithmetic error due to a misplaced decimal point), through mistakes (like misunderstanding a theory), to complex social organizational mix ups with unintended and unwanted consequences. For example insufficient resources may put pressure on the planning process and lead to poor supervision and poor workmanship. Engineers have therefore to ensure proper quality assurance and control. Unfortunately, sometimes safety is threatened by criminal or corrupt violations—so structural engineers need a strong sense of ethics.

Making structures that are safe is a top priority but there are many other requirements too. For example all engineering activity is a business and that imposes significant constraints. The stated needs of a client are inevitably the driving force and a starting point—but often a considerable dialogue between client and designers is required to specify how they may best be met. Projects inevitably involve lots of people with lots of differing needs—it is a team effort. For example the accountant and banker financing a building project will want to know what the costs are before lending sufficient money. The marketing manager will be anxious to get lettable floor space as soon as possible and the developer will see every day as a cost on her return. The surveyors will sort out the use and tenure of land, and keep a track of costs and materials. Project managers will try to ensure that the actual work progresses as planned, including the role of the building services engineers who will design, build, and maintain heating, lighting, and ventilation systems. Throughout the whole project the architect will be concerned with controlling the integrity of her vision.

Costs are always important and often critical. Every project has to be justified through a business case. A business case is a statement or document that presents the arguments for investing in a project and a business plan. The plan contains the strategic intent of the team or organization with formal statements of the objectives, actions, and inputs needed to achieve the stated goals. Within this plan often it is not absolute cost that is the major issue but rather its affordability. Engineers who have an appreciation of how their client is financed will appreciate when the financial arrangements are sensitive to the predictability of a cost. Unforeseen problems can escalate costs and make projects become unaffordable. Alternatively, an inadequate business case can lead to problems. For example, in the past the first cost of designing and building for a developer was too often the focus of attention. The first cost is simply the developer's cost of getting the building built and selling it on to someone else who will own, use, or maintain it. Emphasis on first cost means insufficient attention to sustainable costs and benefits in the full life cycle. Now structural engineers have to consider how to make their structures sustainable—for example by minimizing embodied energy and carbon emissions.

Other, less directly expressed, needs that the structural engineer must consider are from those people who are not part of the project but who have an interest in the success of the project—the stakeholders. These include relevant company shareholders, financial institutions, and, most importantly, the end users— usually, you and I, the general public. Stakeholders do not normally have a direct voice in the project team but may have an impact, for example through lobby pressure groups such as environmental campaigners who protest about the route of a new road or railway. Of course successful clients and the project teams always have the needs and requirements of the ultimate end user in mind when they are drawing up their specifications. For example a cruise ship that not enough people will choose to sail in, or an aeroplane that not enough people will choose to fly in, or a bridge that no-one will cross are of no use to anyone.

Our second perspective is materials—*what* will the structure be made of? The list of materials from which traditional structures are made is actually quite short. It includes timber, masonry, concrete, iron, steel, aluminum, plastics and composites. The science of 'Strength of Materials' or 'Mechanics of Materials' deals with how materials deform under imposed forces and is an important topic in its own right. Our coverage in Chapters 4, 5, and 6 will just touch the surface. The science of new materials is currently developing rapidly so that new alloys of aluminum and titanium as well as plastics and composites such as carbon fibres are being used for aircraft. The A350 Airbus has more than 50 per cent composites and there are currently major developments in the development of new materials that will become important in the years ahead. The specification of materials is one area where structural requirements do vary greatly. For example weight is a premium in aircraft. Weight may not be so critical in ships but its distribution through the length and height of the ship is important for stable buoyancy. Weight can be critical for submarines, hovercraft, and hydrofoils. Bridge and building materials have to be strong enough for the job they will be asked to do, readily available, and not too expensive, since cost is a major consideration. Of course combinations of material are used. For example, because concrete is strong when squashed but weak when pulled, steel bars are used to reinforce it.

The third perspective is, as we have already seen, *what* should be the form of a structure? It is perhaps the most critical decision that a structural engineer has to make and hence is the focus of much of the rest of this book. Michel Virlogeux is a French structural engineer responsible for a number of big bridges including the Millau Viaduct in France. He says that we design beautiful bridges when the flow of forces is logical. The philosopher Ervin Laszlo pointed out that the difference between form and function does not exist in natural structures. In other words the best structures are a harmony of architecture and engineering—where form and function are one. We'll explore that idea more in Chapter 2.

How do forces flow?

First, we need to consider how a force can flow and how that flow can be logical. At its simplest force is a pull (tension) or a push (compression). Take a piece of string and pull on one end then the string will simply move. If someone else pulls on the other end to counter our pull then the string becomes taut as it transmits the two forces—our pull which we call an action and the force at the other end, which we call a reaction. If the action and reaction forces balance so that there is no net movement then we have equilibrium. The question then arises—does the string have to be strong enough to take twice the pull? The answer is no and a little thought experiment will help us to see why. Imagine what would happen if we cut the taut string with scissors. Both cut ends would become free and the strings would move. Now imagine asking someone to take the place of the cut fibres of the string by pulling on each end at the same time. Now we have two pieces of string both in equilibrium. The person has effectively replaced the cut fibres and is doing what they were doing before we cut them. She is providing an internal force that balances the external forces of action and reaction.

How do we interpret this more generally? When external forces are applied to a structure (action) they will be resisted (reaction) and connecting the two groups of forces are sets of internal forces as shown in Figure 3a. All of the forces are in balance or equilibrium. However, we could also think of Figure 3a in a different way. Imagine we had cut the string in two places so that the line represents not the total length of string but just a small piece of it. The diagram would look exactly the same but the forces labelled 'external' would now be internal, and the diagram would now show balanced internal forces—in effect internal forces flowing along the string. This idea of cutting a piece of the structure out and finding the internal forces is an important one that we will use many times.

Just as a piece of string would snap if we pulled too hard on it, a part of a structure may break if the internal force is too large. There are three ways in which materials are strong in different combinations—pulling (tension), pushing (compression), and

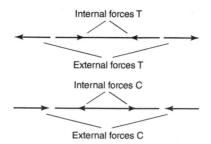

3a. Internal and external forces

sliding (shear). Each is very important, so let's look now at them in turn.

We will do this with the help of a few more experiments we can either think through as a thought experiment or you can do yourself at home. When we pulled on the string there was a resistance—engineers say it has 'tensile stiffness'. Stiffness is simply the amount of force that creates a given stretch on a given length of string. However, if we push on the string there is no resistance—no 'compressive stiffness'.

Now repeat the pulling and pushing on a pencil—we find it has both tensile and compressive stiffness. This is analogous to a steel or concrete column that supports the floors of a building. We will discuss later how engineers calculate the forces actually in the string and the pencil, and how big the forces need to be to break the string or the pencil. For now let's take a piece of paper about the size of a typical bus ticket—say, a 5 cm square. Hold it on two edges and try pulling and pushing again. It will feel exactly as did the string—however, as we compress the paper we will find that it buckles in a curve.

The next thing to try is to move our hands in the plane of the paper as in a slicing or shearing action—we will feel some resistance. The

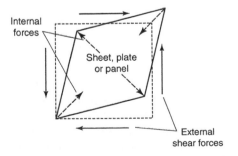

3b. Shear force

paper has stiffness in the plane of the paper—we call that shear stiffness. We will notice a crinkle or buckle forming diagonally across the paper. We can see why in Figure 3b—as we shear the paper we are making one diagonal longer and hence pulling it in tension. At the same time we are making the other diagonal shorter and hence compressing it—and because it has little compressive stiffness it buckles. By contrast if we now try to shear the paper out of the plane of the paper by moving our hands up and down at right angles to the paper then again there will be almost no resistance. In other words the paper has shear stiffness only in the plane of the paper.

So now we have established that string and paper are quite strong and stiff in tension and in-plane shear, but that they are weak in compression. How can we make the paper stronger in compression? If we fold the paper into a tube and press on the ends we will find it much stiffer and stronger. Eventually it will buckle but the nature of the buckle will depend on the length of the tube. If we used a rectangular piece of paper and formed a much longer tube so that the length of the tube compared to the diameter of the tube is much greater, then the tube will buckle into a single curve. A much shorter stockier tube will crumple. Later we will see that the strength of a piece of structure like a tube is quite complex and depends on a number of factors such as how the ends are held and how straight it is to start with.

Now let's look at bending. Take another bus ticket or till receipt, and hold it flat, balanced between two supports such as the forefingers of your left and right hand. Then get someone to press down in the middle—we will see that the paper gives way easily—we say that it has no bending stiffness. Now fold the paper into several strips so that it makes a zig-zag profile. Repeat the pushing down exercise. Then try squeezing it along its length. In both cases we will see how much stiffer the paper is. Simply by folding we have given it lateral and compressive stiffness. For now let's just try one last experiment. For this we will need a cardboard

tube such as the inside of a paper towel we might buy in the
supermarket. Try compressing the ends—it is very stiff and strong
as we might expect. Try twisting it too—it is also very strong in
torsion. Now cut it along its length and do the same. As we might
expect it is much weaker because the cut sides can move relative
to each other. The reason is that before the cut there was an
internal shear force acting along the length between the two cut
sides and stopping them sliding relative to each other. By cutting
the tube we removed its capacity to resist that shear. In Figure 3c
we can see as we twist the tube it warps along its length. We will
find that this kind of warping is very important in large structures
such as ships and jumbo jets.

In summary there are four key points to take from this chapter.
First, the best structures are a harmony of architecture and
engineering. Architectural form and structural form are such that

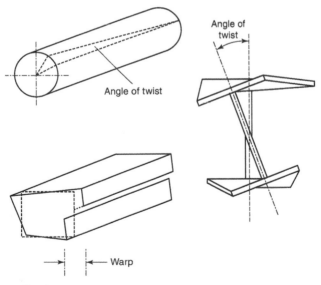

3c. Torsion

form and function are unified and the flow of forces is logical. Second, purpose, material, and form are important criteria in ensuring that a structure meets its specification—including safety and affordability. Third, we can use 'thought-experiments' in which we cut out a piece of the structure out to find the balancing internal forces. Fourth, the strength and stiffness of a structure depends on its form and we can explore and develop our own understanding through some very simple experiments that we can try home.

Finally, as long ago as about 15 BC the Roman Vitruvius wrote that a good building should satisfy three requirements, *firmitas*, *utilitas*, and *venustas*, which translate as durability (firmness, strength, safety, and sustainability), utility (function, commodity), and beauty (aesthetic, delight). But does form follow function or vice versa? That is the question we shall turn to next.

Chapter 2
Does form follow function?

'Form follows function' is another way of saying that the ends determine the means. The phrase was widely used by two famous art historians, Ernst Gombrich (1909–2001) and Horst Janson (1913–82), as a common-sense way of expressing the idea that the shape and look of something should reflect what it is to be used for—a bank should look like a bank and a church like a church.

However, the original meaning, by the American architect Louis Sullivan in 1896, was an expression of a natural law. He wrote, 'Whether it be the sweeping eagle in his flight or the open apple-blossom, the toiling work horse...form *ever* follows function, and this is the *law*...'

It seems that Sullivan was relating the natural form of something to its essential nature—what an object is and what it does. He was suggesting that the natural form is unique, designed by God or by nature with a single solution. The human designer becomes a 'godly messenger', who has to give up her own as well as her client's objectives and become a vehicle through which the true expression of the object can be found.

This wasn't of much practical help to designers. By the mid-1900s it was abandoned in favour of a more common-sense interpretation—but that also soon ran into difficulties because all objects with the same use don't have to look the same.

Because a structure rarely has only one function it is helpful to think of a wider sense of 'meeting a purpose' as we saw in the last chapter. We also noted that Vitruvius asked for *firmitas, utilitas,* and *venustas*—durability, utility, and beauty. A modern interpretation might be resilience, purpose, and delight—indeed *venustas* is a Latin term for the qualities of the goddess Venus and so has a sensual meaning too. Vitruvius saw that a structure wasn't just about utility or usefulness—he wanted more. We will look at resilience in more detail in Chapter 6—in this chapter we will focus on the other two attributes.

In economics, utility is the capacity of a commodity or service to satisfy a human want or need. It can sometimes imply use at the expense of quality. Utility is what many people think of as function but it is only part of delivering a purpose. Function is often interpreted as a question of 'how'—how will the structure do its job? Purpose, on the other hand, is more clearly an expression of 'why'—why do we need this structure? What is it for? Purpose is the starting point. A railway is much more than a track with trains—it generates connectivity, trade, and new developments. But a structure also impacts on the environment through how it looks, how we react to it emotionally (aesthetics), how it uses resources (sustainability of energy and materials), and how it adapts to future needs and behaves when threatened (e.g. by flooding (resilience)). These, and many other interdependent requirements, have to be satisfied for a structure to be deemed fit for purpose. However, before we abandon the idea that form can follow from function there is one aspect that we need to look at more closely—energy.

Later in this chapter we will look in more detail at how a structure contains energy. For the moment you may be able to appreciate its existence by thinking of what happens when a building collapses into a pile of rubble—you witness a massive release of the energy locked up inside the structure.

Structures are naturally lazy

As we said in the last chapter all intact structures have internal forces that balance the external forces acting on them. These external forces come from simple self-weight, people standing, sitting, walking, travelling across them in cars, trucks, and trains, and from the environment such as wind, water, and earthquakes. In that state of equilibrium it turns out that structures are naturally lazy—the energy stored in them is a minimum for that shape or form of structure.

Form-finding structures are a special group of buildings that are allowed to find their own shape—subject to certain constraints. There are two classes—in the first, the form-finding process occurs in a model (which may be physical or theoretical) and the structure is scaled up from the model. In the second, the structure is actually built and then allowed to settle into shape. In both cases the structures are self-adjusting in that they move to a position in which the internal forces are in equilibrium and contain minimum energy. It is then very interesting to realize that these forms are also often very pleasing—and in many cases delightful.

Form-finding structures are perhaps the best examples where a harmony between architecture and engineering can lay claim to being public art. A good architect welcomes the engineering technical discipline to create form through structural art and intelligence; and a good engineer welcomes architectural conceptual discipline to create form through aesthetic art and intelligence. Later in this chapter and in Chapter 4 we will examine more closely how the key to structural art is to form pathways for the logical flows of forces.

Perhaps the simplest form-finding structure is a rope stretched between two points and hanging freely like a washing line. The rope hangs in a shape called a catenary. Robert Hooke (1635–1703) saw that it could be used to define the shape of an arch. He stated, rather boldly, 'As hangs the flexible line, so but inverted will stand the rigid arch'. His idea was that the forces in a hanging cable and an arch are the same except that one is in tension and the other is in compression. Form-finding structures are called *funicular*, derived from the Latin word *funiculus*, meaning thin cord or rope—so a funicular form has only axial forces.

If we hang one piece of heavy washing on the washing line then the rope changes shape and it becomes two straight lines (Figure 4a). As we add more pieces the shape changes again. When the washing is closely crammed together the weight is effectively uniformly distributed along the whole horizontal length of the rope and the shape becomes a parabola. Note that although the external forces applied to the rope are vertically down, the internal force is a tension along the sloping length of the rope.

Unfortunately, as the washing line demonstrates, there is a big problem in using self-adjusting structures in practice. The movements under changing loads can make the structures unfit for purpose. For example walking on a rope footbridge can be rather like walking on a tightrope or trampoline. It may be acceptable, if a little wobbly, as a footbridge but would not be suitable as a road bridge because the changes in curvature would make it very difficult to drive over.

By the end of the 18th century Sir Isaac Newton's theories of gravitation were being used to calculate the internal forces in a structure like our washing line. For example we can draw a line of thrust (compression) or a line of pull (tension) as an arrow on a piece of paper. We can do this because a force is a vector. A vector has a magnitude and a direction. We can contrast it with a scalar that only has magnitude. For example a speed of 20 kilometres

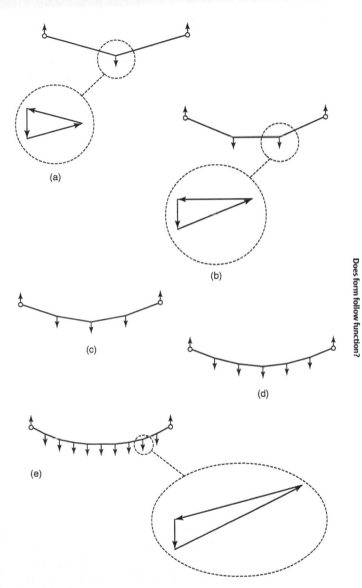

(a)

(b)

(c)

(d)

(e)

4a–e. Hanging rope

(km) per hour is a scalar—it has only a magnitude. A velocity of 20 km per hour heading 5 degrees west of north is a vector—it has magnitude in a direction. The mass of an object is a scalar but its weight is a vector because it is the force of attraction to the Earth and points vertically downwards. The modern international system of units (SI) unit of force is named in honour of Newton—one Newton (N) is the force that produces an acceleration of one metre (m) per second per second when applied to a mass of one kilogram (kg). Therefore because the acceleration due to gravity on the surface of the Earth is 9.81 m per second per second, a mass of one kg weighs 9.81 N (approximately 10 N). We can represent a vector by an arrow drawn in the same direction on a piece of paper using a scaled length. So we can represent a weight of 10 kN (about a tonne force) with an arrow of say 5 centimetres (cm) pointing down the paper. We can also represent other non-vertical forces as vectors too.

These ideas led to methods for calculating forces called 'graphical statics'. We can represent the forces on a washing line by the vector arrows in Figure 4. If the three forces are in equilibrium then they balance and make a closed triangle.

Triangles are important in structural engineering because they are the simplest stable form of structure and you see them in all kinds of structures—whether form-finding or not. Figure 5a shows a two-dimensional triangular pin jointed truss. The three members have only to resist axial tension or compression and have only to be pinned together. Other forms of pin jointed structure, such as a rectangle, will deform in shear as a mechanism (Figure 5b) unless it has diagonal bracing—making it triangular. In Figure 5c the joints are rigidly connected rather than pinned which simply means that the members cannot easily rotate with respect to each other. When this happens the internal angles do not change easily and the adjoining members bend. In other words bending occurs in part of a structure when the forces acting on it tend to make it turn or rotate—but it is constrained or prevented from turning

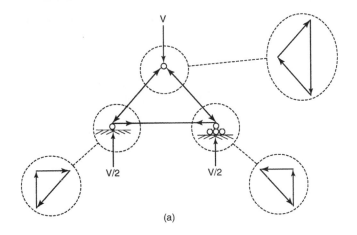

(a)

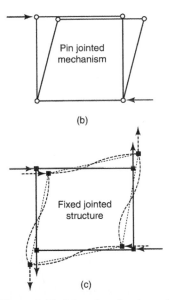

(b)

(c)

5. a Triangle of forces; b Pin jointed mechanism; c Fixed jointed structure

freely by the way it is connected to the rest of the structure or to its foundations. The turning forces may be internal or external.

One of the first important champions of modern form-finding was Antoni Gaudí (1852–1926). He was a Spanish Catalan architect—a modernist—with a highly individual style. He learned about graphical statics as a student and was the first to use them as a way of integrating structural design into architectural design. His best known project is La Sagrada Familia in Barcelona. Starting in 1883 at the age of 31 he worked on it until he died in 1926; it was eventually consecrated by Pope Benedict XVI in 2010. In 2013 work is still going on using modern methods such as 'finite element analysis' (Chapter 4) and modern materials such as reinforced concrete. However, Gaudí was a master builder of the kind that was a precursor to the separate roles of the modern architect and engineer, as we will see in Chapter 3.

Shells rely on shape

After Gaudí, Swiss engineer Heinz Isler (1926–2009) designed over 1,000 innovative reinforced concrete shell roofs using form-finding models. For example one of his simpler methods of finding the form of a shell roof was to hang a wet cloth and let it freeze outside in winter. Another was to use an inflated rubber membrane covered in wet plaster to form a domed surface that could be scaled up. Although there were errors in the scaling up process, Isler was very successful and able to create a kind of structural 'clarity' by omitting the stiffeners that are often needed to prevent the edges buckling when forces are not totally funicular.

Frei Otto (1925–) also used hanging chain nets. He claimed that his structures, such as the German Pavilion in Montreal (1967), the Olympic Park in Munich (1972), and the Multihalle in Mannheim (1975) were 'natural' structures. But he wasn't using the word natural in an analogy with nature but rather that the shapes were obtained through a self-generating 'natural' form-finding process. In other words he was referring to the idea that the form was not decided by the artistic whim of the architect but by the self-shaping process obtained using a physical model.

Modern theory tells us that these pioneers were relying on the idea of a 'minimal' membrane with equal stress in every direction in the plane of the surface. One way of imagining this is to think of a soap film or bubble made from a child's toy soap bubble wand. Within a given boundary or set of constraints the bubble will have the smallest area possible. A synclastic bubble is one where the curvature in two perpendicular directions is of the same sign, in other words convex or concave—double curvature, like a sphere.

By contrast a surface with an anticlastic double curvature is a saddle surface. A key feature of a saddle surface is that at any point on that surface and in any direction, the curvature of the surface will be of the opposite sign to the curvature at that same point but in a perpendicular direction. In other words if the curvature in one direction is convex then at 90 degrees the curvature will be concave. Such a shape we now know is stable.

In 1973 the London-based structural engineering company, Ove Arup, were consulted about the initial design by Frei Otto for a grid shell for the Multihalle Pavilion in Mannheim, Germany. A grid shell is simply a shell with regular holes—rather like the wire mesh of a kitchen sieve. The shape was found by inverting a net hanging under its own weight. Otto proposed a grid of timber laths approximately at right angles and tightly connected at the joints so that each rectangle can resist shear (Figure 5c). The Arup team, led by Ted Happold, concluded that the initial designs, based on physical models of the type used by Gaudí and Isler, required strengthening. The final Arup design consisted of two rectangular grid shells each made up of wooden laths of only 5 cm by 5 cm and covering a maximum span of 60 m. The two sets of laths were laid flat on the ground and loosely bolted at the joints by a single bolt. The whole structure was then lifted over a scaffold and allowed to settle into its natural shape. The final form was locked-in by tightening all of the bolts and fitting diagonal cables across many of the grid holes. The effect was rather like covering a football with a net curtain without wrinkles. At the top of the ball the net would be square but over the sides the net must shear into an increasingly acute diamond shape.

A more modern example built on the same principle is the Downland Gridshell designed by Buro Happold and built for the Weald and Downland Open Air Museum in West Sussex and opened in 2002. The grid consists of timber oak laths although this time the shape was determined geometrically rather than as an upside-down hanging net. The oak laths were 'finger-jointed' together into standard lengths of 6 m which were then joined to form laths of 36 m in length. As at the Multihalle the two grids were laid out flat on the ground, loosely connected by bolts, and lifted onto a supporting scaffold using cranes. The two grid layers slid over each other as the structure slowly deformed and took its own shape. When the full shell was formed the edges were secured to a timber platform and the bolts at each node were tightened. In-plane shear stiffness was provided by large timber shingles nailed to wooden stringers running diagonally across the shell. The reason for two grid layers was to resist loads other than self-weight—for example when it snows the line of thrust changes but the designers had to ensure that it still would lie within the depth of the grid.

These kinds of rather exciting form-finding structures are part of a more general class of structures called 'shells'. Shells can be solid like an egg shell and Isler's reinforced concrete roofs. They can be grid or lattice nets like a kitchen sieve or the Downland Gridshell.

Unfortunately if the shape of a shell structure isn't quite right there can be major problems such as ponding of rainwater and snow, buckling of edges, and the creasing and tearing of fabrics. Indeed a key challenge for fabric structures is to design a suitable cutting pattern—just as a clothes dress designer has to create a cutting pattern that ensures a ladies' dress will hang properly.

The impressive Eden Project in Cornwall, England (Figure 6), opened in 2001, has two giant greenhouses—each one made of four intersecting spherical geodesic domes. These are called biomes because they host typical vegetation of different climate zones of the world. The biggest biome has a diameter of almost 125 m and a free height inside of 55 m. The structure is in two layers with an outer hexagonal grid that is connected by diagonal elements to an inner tension net. This structural depth provides resistance to loads from snow drifts and slides, and wind. The cladding consists of air-filled Ethylene tetrafluoroethylene (ETFE—a fluorine based plastic) cushions because they are much lighter than glass and allow much more ultra-violet light through

6. **The Eden Project**

with good heat insulation. Truss girders take the loads at the intersection of the biomes—the biggest one has a span of 100 m.

Another class of interesting and rather special structures are called 'tensegrity structures'. They have isolated axial compression members suspended by a continuous network of tension cables as shown in the model in Figure 7. They have a very light and almost ethereal appearance as the more noticeable compression members appear almost to float against the almost imperceptible tension cables. Buckminster Fuller coined the name 'tensegrity' from 'tensile integrity' as the integrity or stability of the system comes from the tension members. David Geiger designed two tensegrity cable domes for the 1988 Olympics in Seoul, Korea, and others have followed. A cable-net structure, like the London Eye, uses tension wires just as they are used in a bicycle wheel. Pre-tensioned spokes pull on the hub putting it into tension and pull out on the outer rim putting it into hoop compression. The roofs of the London 2012 Olympic Stadium and the Cape Town Stadium

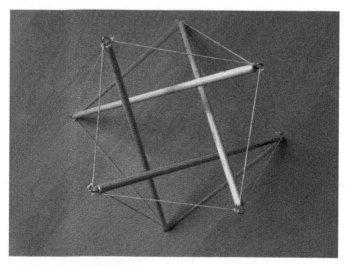

7. Four bar tensegrity model

8. a Cape Town Stadium; b Cape Town Stadium roof

(Figure 8) work in a similar way as horizontal bicycle wheels and an open tension ring for the wheel hub.

If the lengths of the tension cables are adjustable, say by using a turnbuckle, the form of a tensegrity structure can be varied. This makes them suitable as deployable structures—rather like the unfolding of an umbrella. For example they can be folded, sent up by rocket into space, and then unfolded to form a space structure. Deployability is attractive when pre-use storage space is at a premium or for compact transportation. They can even be self-erecting by using the energy in the elastic cables to deploy the structure into its stable configuration. Tension members also self-stabilize and become more stable the larger the load, as the tension pulls the structure towards the equilibrium position.

We said earlier that a rope hanging freely as a catenary has minimum energy and that it can only resist one kind of force—tension. Engineers say that it has one degree of freedom. Let's now look at these ideas more closely since they form one of the first parts of our grammar of structural theory and will help us to identify the paths of the flows of forces.

Force pathways are degrees of freedom

Energy is the capacity of a force to do work. If you stretch an elastic band it has an internal tension force resisting your pull. If you let go of one end the band will recoil and could inflict a sharp sting on your other hand. The internal force has energy or the capacity to do work because you stretched it. Before you let go the energy was potential; after you let go the energy became kinetic. Potential energy is the capacity to do work because of the position of something—in this case because you pulled the two ends of the band apart. In a bridge structure the two ends of a piece of steel or concrete might move apart due to an internal force. A car at the top of a hill has the potential energy to roll down the hill if the brakes are released. The potential energy in the elastic band and

in a structure has a specific name—it is called 'strain energy'.
Kinetic energy is due to movement, so when you let go of the band
or if you release the brake the car will roll down the hill and the
potential energy is converted into kinetic energy. Kinetic energy
depends on mass and velocity—so a truck can develop more
kinetic energy than a small car.

When a structure is loaded by a force then the structure moves in
whatever way it can to 'get out of the way'. If it can move freely it
will do—just as if you push a car with the handbrake off it will roll
forward. However, if the handbrake is on the car will not move,
and an internal force will be set up between the point at which you
are pushing and the wheels as they grip the road. Under gravity all
structures, like the freely hanging rope, settle towards the surface
of the Earth as far as they can. So if we think about one small
piece of the rope in the catenary, that piece will fall as far as it can
until it is constrained by its connection to the rope on either side
of it. This is why it has minimum potential energy. It is just as a
climber, slipping off a mountain ledge but roped to another
climber, would not fall all the way down but would be held by the
rope, as long as the other climbers could hold on.

The rope in tension has one degree of freedom, one pathway for
the flow of force, because it cannot resist any other kind of force. In
brief, degrees of freedom are the independent directions in which a
structure or any part of a structure can move or deform—the force
flow pathways that are the key to structural art. We'll come back to
the important word, 'independent', a little later. Movements along
degrees of freedom define the shape and location of any object at a
given time. Each part, each piece of a physical structure whatever
its size is a physical object embedded in and connected to other
objects—just as we said a piece of rope is connected to the adjacent
rope on either side. Whether the object is a small element of a cross
section 1 millimetre (mm) by 1 mm or a large sub-structure like the
deck of a bridge, it is connected into other similar objects which
I will call its neighbours. Whatever its size each has the potential to

33

move unless something stops it. Where it may move freely—as the car rolling down the hill with no brakes on—then no internal resisting force is created. However, where it is prevented from moving in any direction a reaction force is created with consequential internal forces in the structure. For example at a support to a bridge, where the whole bridge is normally stopped from moving vertically, then an external vertical reaction force develops which must be resisted by a set of internal forces that will depend on the form of the bridge.

So inside the bridge structure each piece, however small or large, will move—but not freely. The neighbouring objects will get in the way rather like people jostling in a crowd. When this happens internal forces are created as the objects bump up against each other and we represent or model those forces along the pathways which are the local degrees of freedom. The structure has to be strong enough to resist these internal forces along these pathways. Note that our representation, our model, is not the reality. It is a description of the reality for a specific purpose and from a specific point of view. A model is inherently incomplete in a way that we will see is important when later we think about risk (Chapter 6).

The degrees of freedom are defined using a mathematical convention. A rigid body like a brick has six degrees of freedom which are defined locally, in other words relative to its particular orientation. So it can move along three axes at right angles (x, y, z) where x is not necessarily horizontal but along the length of the body. It can also rotate about each axis (Figure 9) labelled as $(z\,y)$ or rotation in the $(z\,y)$ plane about the x axis, rotation in the $(z\,x)$ plane about the y axis and in the $(y\,x)$ plane about the z axis. As I said earlier, these directions must be independent. What this means is that it must be possible for a change in one direction without a change in the others and that is why the axes are at right angles or 90 degrees to each other. Without independence it would not be possible to separate out the various movements and the consequential internal forces. Nor would it be possible to

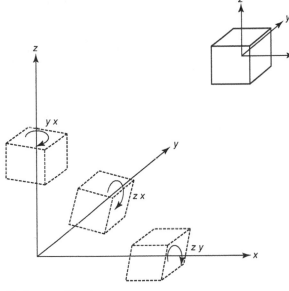

9. Six degrees of freedom

combine them again. Of course in general they all do change together. By describing those changes in terms of the chosen independent degrees of freedom the structural engineers can make sure that each little piece of the bridge is able to resist the internal forces along these six directions. The grammar of the theory helps structural engineers to understand the nature of these internal forces.

You may remember in Chapter 1 and Figure 3a we identified the balanced external and internal tension forces on a piece of string (rope or cable) that we represented by a single line along one degree of freedom. In Figure 10 we now look at the forces on a small block of material in one, two, and three dimensions. In Figure 10a, like Figure 3a, the external force is an axial tension on one axis—it is uniaxial. In Figure 10b there is tension on two axes—it is biaxial and in Figure 10c we have triaxial tension. Of course if any of these forces were compressive then the arrows would be reversed. Under these forces the block will deform—stretch or compress—and we can capture the change by describing it as movement along each of the three degrees of freedom (x, y, z). In Figures 10d–f we show what happens when the block is sheared and bent. As we said in Chapter 1 the vertical shear forces act along the vertical sides of the block but because they are separated they tend to turn or rotate it. This tendency is resisted by horizontal shear forces. Other forces that tend to rotate the block are called bending moments—shown by a curved arrow in Figure 10d. Note that a moment here does not refer to time but to rotation. The bending moment action force on one face of the element is resisted by a reaction bending moment on the opposite face.

Figure 10g shows the direct, shear, and bending forces in three dimensions and Figure 10h shows how the turning forces or moments can produce bending and twisting. These diagrams are quite complicated. On top of this, different members of a structure are aligned in different directions and so the (x, y, z) axes for two or more members will, in general not be the same. Engineers therefore normally translate the local axes for each individual member into global axes in which x and y are horizontal and z is vertical. It is important to realize that degrees of freedom may be defined in these global axes for the whole structure but it is easier to understand the flow of forces along the local axes of each member.

Until the advent of powerful computer programs in the last part of the 20th century all but three of the degrees of freedom were dealt with by common-sense because it was too complicated to calculate them. So, for example, two bridge beams or trusses were placed side by side and braced together. This stabilized them sideways or laterally, and prevented twisting. It was an intuitive and effective way of dealing with all of the degrees of freedom that have a sideways lateral component. As a consequence the focus for many

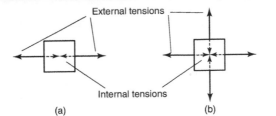

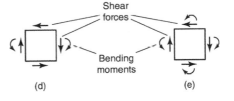

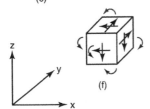

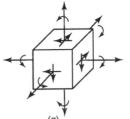

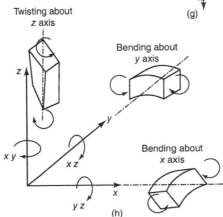

10. a–g Forces on elements; h Bending and twisting forces

centuries of engineering scientific development was on just three of the possible six degrees of freedom. The first was the vertical movement which contributes to the displacement of the bridge. The second was the horizontal movement along the length of the bridge which contributes to its change of length. The third was rotation in the vertical plane as you look at the side view or elevation of the bridge, which when resisted creates internal bending moments. In Chapter 4 we will examine how modern engineering science and, in particular, extremely fast computer calculations have enabled us to develop 'finite element' techniques of theoretical analysis through which we can understand, calculate, and control all six degrees of freedom for every part of the structure. As we have seen already this knowledge is enabling us to build much more exciting and unusual structural forms.

In summary there are four key points to take from this chapter. First, when a structure is loaded by a force it moves in whatever way it can to 'get out of the way' and settle towards the surface of the Earth as far as it can. In other words structures are naturally lazy—they contain minimum potential energy. Second, each piece of structure, however small or large, will move—but not freely. The neighbouring pieces will get in the way. When this happens internal forces are created as the pieces bump up against each other. Third, we represent those forces along the pathways and call them degrees of freedom. Finally, the structure has to be strong enough to resist these internal forces along these pathways.

Of course whilst these exciting and interesting form-finding structures are examples of the fusion of engineering and architecture they are relatively few in total compared to the buildings we see around us. So we now need to turn to the many more other types of structure—but first a little history.

Chapter 3
From Stonehenge to skyscrapers

An understanding of the past is an important influence on the way we approach the future. Through the past we learn for the future. The purpose of this chapter is to generate a sense of history of structure as part of human progression—not to describe lots of factual detail. I shall argue that one of the challenges of the future is to get back to some of the more integrated ways of the past.

We'll divide this brief review into four periods—the ancients to the late middle ages and early Renaissance; the Renaissance to the early 18th century; around 1700 and the beginnings of modern times up to about 1900; and then finally the last century up to the present.

One of the earliest major structures in Europe was Stonehenge in Wiltshire, England, built from around 3100 BC and a ring of massive standing stones set within earthworks and burial grounds. Why it was built is not clear—there are many different theories ranging from its being a site for religious ceremony to its being a gigantic solar calendar. However, Stonehenge demonstrates very clearly that we humans can achieve the extraordinary. The pyramids in Egypt were equally impressive ancient structures dating from about 2600 BC. We know that the Egyptians developed elaborate beliefs about life after death. For example a pharaoh's dead body had to be kept intact in order to proceed to the afterlife—hence the practice of mummifying bodies and the building of large tombs. We have no knowledge of the architects and engineers at Stonehenge, but men like Khufu, Hemiunu, and Imhotep are credited by Egyptologists as the men who built some of the pyramids. The first burial chambers, called *mastabas*, were square buildings with a room inside for the coffin, mummy, and items to take to the afterlife. The steps in the pyramids were made up of successively smaller *mastabas* built on top of each other. Eventually the steps were filled in to create the classic pointed pyramid such as Khufu's Great Pyramid at Giza. As with Stonehenge, many people have speculated about how the work was done. For example a recent proposal about how the blocks were lifted to the top of the pyramid is that the lower blocks were hauled up an external ramp whilst a second internal spiral ramp was built and used for the top two-thirds.

There may be a tendency to think that large-scale engineering failure is a relatively recent phenomenon. However, the pyramid at Meidum tells us otherwise. It was built in three distinct stages—first, a seven step pyramid; second, an eight step pyramid over it; and, finally, an outer mantle making a true pyramid. However, only the lowest part now remains because there was a massive collapse—it seems that there were major design faults. The supporting buttress walls were fewer and more widely spaced than for any previous pyramid. The outer masonry was only anchored by a single layer of mortar to smoothed surfaces, and was founded on sand rather than the rock used below the stepped pyramids. Fourth, the outer packing blocks were not well squared, resulting in an outward force which rose steadily with height.

So what do these ancient structures tell us? First, the failure at Meidum shows us that attention to detail is important. Second, with strong motivation and good organization (the workers were not slaves) people can achieve the extraordinary. Third, modern experimental attempts to replicate what was done have demonstrated that brute force was not sufficient. Within the knowledge and experience of the time the builders clearly exhibited great ingenuity. They were engineers in the true meaning of the Latin root of the word *ingeniarius*—someone who is ingenious in solving practical problems.

Basic human needs for more mundane structures in these early times were much the same as they are now. For example, then as now, structures were made for shelter (buildings), worship (churches and temples), water storage (dams), and travelling over gaps (bridges) and water (ships). The first known boats around 10,000 years ago were more like rafts than ships, but by 3000 BC the Egyptians had assembled wooden planks to make ships with hulls. An example of this is the Khufu ship, which had been entombed at the foot of the Great Pyramid at Giza and found intact in 1954. The Minoans, Phoenicians, Greeks, and Chinese were amongst the early naval powers.

Of course our needs are greater today—to build cars, trains, and aeroplanes as well as power stations, oil rigs, stadia, and other structures. However, then as now, two factors are dominantly important. The first is the *thinking* behind what is required. What is the need or problem? What are the potential solutions? Then designing or deciding what to make. The second is the actual *making*—the manual work to create the solution. Although the required skills may be different, they are totally interdependent. The second relies on the first because if the thinking is poor then the making will be all the more difficult. But the first also relies on the second because if the thinking doesn't consider how something will be made it will almost certainly not work as it should. Throughout history the balance of emphasis between these two factors has not been straightforward.

For example the ancient Greeks are remembered as philosophers and thinkers rather than as builders and makers. Plato had enormous influence but not entirely for the best. He maintained that perfection could only come from the mind and so practical action was inferior. Euclid was the first to formalize geometry by defining axioms and deducing consequences. Archimedes followed Euclid's methods by trying to deduce laws of mechanics and his famous principle of hydrostatics. The oldest known textbook on mechanics was written by Aristotle or his pupil Straton of Lampsakos with sections on gear wheels, levers, breaking strength of wood, galley oars, and an explanation of how boats could sail into the wind.

By contrast the Romans are remembered as great builders and makers but not great philosophers and thinkers. For example they used the arch to great effect but they didn't need any rules of proportion because all of their arches were semi-circular. Magnificent examples of their work are the aqueduct, the Pont du Gard in France, the Colosseum in Rome, and Hadrian's Wall which once divided England and Scotland. Another Roman masterpiece is the Pantheon in Rome, built in 120 AD. The spherical 44 m diameter dome was built of concrete with lightweight aggregate with a density and thickness that reduces towards the crown. The dome rests on walls 6 m thick and 30 m high which encompass chapels that effectively move the centroid of the walls outwards and help to resist thrust. It is a beautiful and sophisticated building that demonstrates the integration of architecture and engineering.

One of the first people to write about the thinking behind Roman structure was Vitruvius and his three principles of *firmitatis*, *utilitatis*, and *venustatis*, which we looked at in Chapter 2. Vitruvius used the only theoretical language available to him—geometry—to record rules of thumb for proportioning structures. For example part of a rule for a Roman Forum read, 'The columns of the upper tier should be one fourth smaller than those of the lower ....' Delight or beauty, he wrote, depends on six aspects—some of which depend on geometry. The first, he wrote, is *ordinatio* or being well ordered, arranged, and methodical; then comes *dispositio* or organization or arrangement of the whole; followed by *eurythmia* or good proportion or rhythmic movement; *symmetria* or relations between elements such as basing everything on a unit module; *decor* or etiquette, elegance, beauty, grace or charm; *distributio* or distribution by assigning roles, specifying duties, and organization skills needed to get a building built.

Although Vitruvius would have been familiar with the term 'architect', even as late as 1600 there were no architects in the sense that we understand them today. The roles of engineering and architecture were indistinguishable. Structures, such as the magnificent gothic cathedrals with their arches, vaults, and flying buttresses, were created by many people together—described loosely and often interchangeably as architects, masons, builders, and carpenters. Individuals took the lead as seemed right at the time and drove the building process. Some became known as master masons and architects.

Introducing the artist engineers

By the start of our second period of history in the later middle ages, major Western buildings were being built by craftsmen plying their separate trades—mainly masonry and carpentry. Some by virtue of skills and personality were natural leaders. Historians tend to refer to them as the architects—though that is not what they were called in the documents of the time, so the title is misleading. Their training and experience was such that they developed from practised craftsmen into 'master builders'. The term 'architect' at this time sometimes refers to a technician and sometimes to patrons or clients who often did their own design work. During the Renaissance patrons in Italy who wanted monumental buildings started to look to artists for help. For example Alberti (1404–72), Bramante (1444–1514), Brunellesco (1377–1446), Leonardo (1452–1519), Raphael (1483–1520), and Michelangelo (1475–1564) were typical Renaissance polymaths working as artists, poets, philosophers, architects, and engineers.

Brunellesco was responsible for the Duomo in Florence; Bramante, Raphael, and Michelangelo were each at various stages architects for St Peter's in Rome. The Duke of Milan appointed Leonardo da Vinci as *Ingenarius Ducalis* (the Duke Master of Ingenious Devices). As a consequence respect for craft skills began to decline. Conceiving, drawing, and *disegno* (Italian for drawing and design) began to hold sway with scholars and princes who increasingly controlled the flow of information through printed books. It hardly matters if we call the great designers of the Italian Renaissance artists, architects, or engineers—at the time the terminology related to the job not the person. However, it was from this time that the term 'architect' came to be associated with artist and conceptual thinker—but it was a commercial art form set in a business context.

In Italy and to a lesser extent in France, the authority of the architect began to assert itself in the 16th and 17th centuries. Palladio (1508–80) wrote four books of architecture published in 1570 and his influence was imported into England by Inigo Jones (1573–1652). In Tudor and Stuart Britain medieval practice was still the norm—Elizabethan houses were designed piecemeal rather than as artistically coherent entities. Craftsman architects still dominated the scene at the beginning of the 17th century. Again change came from patrons with changing tastes. Jones, who was surveyor of the Royal Works for over 20 years until the start of the English civil war in 1642, designed the Banqueting House at Whitehall and the Queen's House at Greenwich. He imposed an Italian discipline on English architecture which was beyond the grasp of the average master builder. Consequently the craftsmen became subordinated to the designer—a single controlling mind—in a manner never seen before.

In 16th century Italy and in the France of the *grand siècle*, people were responding to the power and purpose of monarchy. The *Académie des Beaux Arts* (which became the *Ecoles des Beaux Arts* in 1863) was founded in Paris in 1648 to serve the needs of royalty and government. Some of the workers were in reasonably continuous employment and they began to exchange ideas. Professional groups made up of a sizeable number of experts began to appear. The engineers of the *Corps de Génie* took charge of war and infrastructure, while the architects of the *Bâtiments du Roi* housed the king and helped articulate his magnificence, his generosity, and his piety. In the time of Louis XIV, the Marquis de Vauban (1633–1707) was Marshal of France, a military engineer, and *ingénieur ordinaire du roi*. The techniques of architect and engineer were much the same, so Vauban employed people we now think of as architects, and when the *Corps des Ponts et Chaussées* was set up in 1716, its first chief engineers were *architectes du roi*.

Major structures, such as cathedrals, needed people skilled with machinery such as pulleys, hoists, and ladders. The word *ingeniator* was being used to describe some of the people who were expert in machines, but these were mainly machines of war. For example, in England, Ailnolth (1157–90) was one of the first men to be called an engineer—he was a craftsman with some technical training in the making of the large war engines which preceded the use of gunpowder. As the training for these various jobs varied, so did their skills and talents. Some became specialized, but many moved between fields because they weren't able to spend a whole lifetime in one specialism. As a consequence those who designed secular or religious buildings tended to be called architects but those who designed forts, walls, towns, ports, canals, or machines for war were increasingly referred to as engineers.

In 1775 the French school formed for the *Corps* became the *Ecole des Ponts et Chaussées*. Thus the primary thrust for engineering education came from the military. Nevertheless in Europe as a whole there were growing civilian needs. The first man to call himself a civil engineer—a non-military engineer—was John Smeaton (1724–92).

Up to this time the only theoretical language for practical work was geometry. But progress in science during the Renaissance was soon to change that. For example Simon Stevin (1548–1620) developed the triangle of forces (Chapter 2) and the decimal system. Galileo Galilei (1564–1642) estimated the breaking strength of a timber beam. And, finally, Sir Isaac Newton (1642–1727) developed the laws which became the fundamental theoretical tools of modern engineering.

In summary, in Europe between 1400 and 1700, the architect and the engineer were differentiated more by the hierarchies they belonged to and the tasks they carried out than by the building techniques they used or the design skills they employed. The medieval architect was a master craftsman. Architecture was not a liberal art, but architects had begun to enjoy a certain status.

The separation widens

In our third period of history, from about 1700 to 1900, the processes of separation between architecture and engineering that started in the Renaissance gained momentum. The term 'naval architect' began to be used. Specialisms emerged to meet new needs—for example to exploit new materials such as iron and steel, and the rational methods of the new science. The scientific approach to structural problems of statics and strength began to make it possible to cope with extensive and difficult tasks in more economic and safer ways. So it began to be possible to design structures from two different points of view—the one emphasizing structure and the other the aesthetic appearance. According to the nature of the task one or the other prevailed—engineering for utility buildings and architecture for monumental buildings.

The almost inevitable gap widened in the 19th century. There was a steady demand for construction projects of different types, presenting new problems. People in public employment as well as those in private appointment could now specialize. Naturally, they came together to discuss mutual interests and consequently formed clubs, societies, and professional institutions. By 1840 there were established professions of civil engineering with the Institution of Civil Engineers (founded 1818) and of architecture with the Institute of British Architects (founded 1834). The Royal Institution of Naval Architects was formed in 1860 and the Institute of Marine Engineers in 1889. Naval architects are engineers concerned with making ships structurally safe, with a form that has a low resistance to flow through the water, and which is stable at sea. Marine engineers are usually sea-going and deal with mechanical and electrical equipment—they are sometimes referred to as 'naval engineers' in the navy. Members of the two Institutions are Chartered Engineers in the UK.

At first the new science had little impact on the practical methods of structural engineering. However, by the middle of the 18th century it was beginning to be useful. One of the important early figures in the development of the new science was Charles Coulomb (1736–1806) who is perhaps better known for his discoveries in electricity and magnetism. He was an engineer in the French army and wrote papers on statics, laying the foundations for a theory of soil mechanics which was essential for the design of foundations. Claude-Louis Navier (1785–1836) established structural analysis as a science. Augustin-Louis Cauchy (1789–1857) introduced the idea of stress (Chapter 2). Jean Claude Barré de Saint-Venant (1797–1886) made significant additions to Navier's work. Carl Culmann (1821–81) published the first systematic treatment of graphical statics (Chapter 2). James Clerk Maxwell (1831–79) was the first to analyse statically indeterminate structures (Chapter 4). Christian Otto Mohr (1835–1918) represented stresses at a point now famously known as the Mohr circle. Carlo Alberto Castigliano (1847–84) was one of the first to realize the importance of energy in structures (Chapter 2) and, for his PhD, he developed important theorems later generalized by Freidrich Engesser (1848–1931).

The impact of this scientific progress in structural analysis became greater as iron and steel became available in commercial quantities and new types of structure could be built. By replacing wood with coke, Abraham Darby I was able to smelt iron in substantial quantities from 1709 onwards. His grandson, Abraham Darby III, erected the famous cast iron arch at Ironbridge, Shropshire, over the River Severn in 1777–9, and today this bridge still takes pedestrian traffic. Steel could be made, but not in sufficient quantities. In 1784, Henry Cort produced wrought iron in a coal-fired flame furnace through the so-called puddling process, with the result that the iron was produced faster than the forge could deal with. Cort then invented grooved rollers for making bars and plates which previously had to be hammered and cut from hot strips.

Smeaton was the first engineer to use cast iron to any great extent for windmills, water wheels, and pumps. Following his practice, cast iron beams were I-shaped but with small top (compression) flanges and large bottom (tension) flanges. Although suspension bridges had been built since ancient times, James Finley was the first to build one in Pennsylvania in either 1796 or 1801 with a stiffened deck and iron chains. The Menai Bridge (1823–6) built by Thomas Telford used wrought iron flat links and, because of the paucity of data, he carried out extensive tests. Eaton Hodgkinson (1789–1861) and William Fairbairn (1789–1874) tested the strengths of cast iron, wrought iron, and later steel. The results of these tests were widely used, not only in Britain but also on the continent.

Whilst the French engineers with their college educations were developing the theoretical side of structural engineering, the British were developing the practical side. Hodgkinson and Fairbairn were both sons of farmers; they had little in the way of formal education but both rose to become members of the Royal Society. They developed successful empirical formulae for the design of beams and columns which were widely used in Britain.

In his book *On the Application of Cast and Wrought Iron to Building Purposes*, published in 1870, Fairbairn quoted some of them, but he also reported the tragic failure of cast iron beams at a mill in Oldham in 1844 when 20 people were killed. He gave formulae for the strength of truss beams of the type used for the Dee Bridge which failed in 1847, and discussed the design of tubular bridges such as those at Conway and Menai. In 1843, Isambard Kingdom Brunel built the first ocean-going propeller-driven iron ship the SS *Great Britain*—now housed in Bristol Docks, UK.

In 1855, Henry Bessemer (1813–98) conceived the idea of replacing the traditional, laborious, and costly puddling process with the mechanical process of blowing a blast of air through the fluid pig-iron. This led to the development of steel production in large quantities at economical prices. William Kelly (1811–88) discovered the process at about the same time in Kentucky, but unfortunately he went bankrupt trying to develop it. The first bridges built using the Bessemer steel were in Holland, but the steel was of poor quality and as a result the use of steel was prejudiced for many years. However, the micrographical work done by Robert Hooke in 1665 had laid the foundations for metallurgy and the problems were soon solved.

As we saw earlier the first schools for engineers were founded in France by the military. By contrast, in the maritime countries, Holland and Britain, a spirit of commercial enterprise and overseas trade was allowed to grow and maritime trade flourished. From the 17th century the French navy developed a new type of ship called 'the ship of the line' with broadside guns. The quest for ever greater efficiency and different uses such as fire-fighting, rescue, and research resulted in many new ship designs.

Education in Britain was largely a matter of individual enterprise. Although there were numerous English books and periodicals in circulation in 18th century Britain, John Desaguliers (1683–1744), once an assistant to Newton, criticized the unscientific empiricism of British engineers. However, it could not be denied that they were just as successful as their French counterparts.

In fact the British engineers who led the technological revolution were, first and foremost, practitioners, learning what they could by independent reading. Their main books were notebooks in which they recorded anything they saw which might be useful—they acted as their own researchers. They felt they were dealing with forces and effects which Smeaton described as being 'subject to no calculation' such as rain, wind, waves, etc. Mathematical analysis after the French example was to them a luxury for which they could hardly afford the time; it was the business of mathematicians and scientists, and in any case it was unreliable unless well supported by experience.

Very little was being done in Britain to provide a technical education for the lesser engineers. The Mechanics Institutes were formed, first, in Glasgow in 1824 and, a year later, in London and Liverpool. Many more were established in response to the need for adult education for craftsmen. A magazine, *The Mechanics Magazine*, specifically for 'intelligent mechanics', was eventually replaced by today's magazines *Engineer* and *Engineering*. In 1831, the British Association for the Advancement of Science was formed. The great Exhibition of 1851 was housed in a large glass conservatory. It was characteristic of the English approach that the designer Paxton was neither an engineer nor an architect, but a gardener.

The professions fragment

The final period of our brief summary of the history of structural engineering takes us from about 1900 to the present. During this period the fragmentation of disciplines and the formation of specialisms became a real issue. As we struggle to cope with cross disciplinary issues such as climate change and sustainability, the need for a new integrated approach has become urgent. Providing the much needed depth of specialism at the same time as the breadth of holistic vision requires a systems thinking approach (Chapter 6).

In Britain, at the beginning of this modern period, the idea of independent consulting engineers aloof from actual construction contracts emerged. However, where they worked with architects, they were junior partners. The architect Sir Gilbert Scott (1811–78) seems to have been one of the first architects to regularly use engineers along with other great architects of the period such as Frank Lloyd Wright (1867–1959). The scientific and technical details of steel and reinforced concrete construction was hard for architects to understand. Therefore, after about 1900 any masonry fronted structures with steel frames were 'farmed out' to experts, often after the architectural design was well advanced. In 1908, the UK Concrete Institute was formed—at that time reinforced concrete was a relatively new material. The first members of the Institute were architects, engineers, chemists, manufacturers, and surveyors. In 1912, the Institute embraced all areas of structural engineering, particularly steel frames. Structural engineering was defined as 'that branch of engineering which deals with the scientific design, the construction and erection of structures of all kinds of material'. In 1922, the Institute became the Institution of Structural Engineers.

At the start of the 20th century a new problem began to make itself felt—the need for standardization in products. Sir Joseph Whitworth's favourite illustration was 'Candles and candlesticks are in use in almost every house, and nothing could be more convenient than for candles to fit accurately into the sockets of candlesticks, which at present they seldom or never do!' An Engineering Standards Committee was formed by the various engineering institutions in 1904, and a year later the Government joined in. One of the first tasks was to enquire into the advisability of standardizing rolled iron and steel sections for structures. One of its first publications listed section sizes and quoted some standard formulae such as the deflections of simple beams under various types of loading. The publications of the committee soon became known as British Standards and included specifications for the steel and for standard test pieces. One of the first standards for structural design was published in 1922 for steel girder bridges with specified loads and permissible elastic stresses (Chapter 6).

As the science became ever more powerful so did the specialization. For example in 1903 a new specialization began to emerge after the first successful powered flight by the Wright brothers. The first flimsy, odd looking machine was made of spruce and cloth with two wings, one placed above the other. Aeronautical and aerospace structural engineers are now very highly specialized because minimizing weight is so important; they use aluminium and lightweight composite materials—as we shall see in Chapter 5.

By the end of the 20th century, construction engineering was fragmented into sub-disciplines. The architect is still often described, particularly by the media, as the moving force behind a scheme. In fact, the way the various design professionals interact is governed by the needs of the client, the contractual arrangements, and the nature of the end-product. For example if the client decides on a contractor-led design-build contract then the architects and engineers play a subservient role. Architects are marginal to most infrastructure work while on domestic buildings engineers do little except check foundations and roof structures.

Some architects have produced designs of flagrant imagery for buildings of high prestige. For example the Guggenheim Museum in Bilbao, Spain, architected by Frank Gehry (1929–), has been criticized for its showy, daring, functionless forms that do not fit the locality. Engineering such structures is challenging, with little credit from art critics. Engineering is completely subordinated to architecture. Andrew Saint has written, 'In an art obsessed world, the architect had dragged the engineer out of the temple of reason and beguiled him to worship in the temple of art.'

Clients have a variety of needs, and their decisions, based on their values, resources, and criteria, determine the balance of power in the building team. Almost all structural engineers will provide a structural scheme no matter how architecturally extravagant. The question is then very basic: 'How big is the client's wallet?'

In truth, if there is no coming together between all of the requirements—especially the aesthetic and structural form—then the solution is likely to be unsatisfactory. It is for this reason that architects are now no longer automatically the leaders of the building team.

Large structures are in the public domain, but can they be public art comparable to the fine arts? Architecture and engineering are unquestionably both arts in the sense of the Greek *techne* meaning skill. The skills of the pre-industrial engineer did not differ greatly from those of the architect even when employed on different tasks. If the skills are the same then the art must be the same. Structures can evoke emotions, as do the fine arts, whether designed by an engineer or an architect. If the level of emotion evoked is the same then the level of art must be the same. Historically, and perhaps unfortunately, the advance of science has brought less emphasis on the aesthetic in structural engineering education. Building on the three principles of Vitruvius, David Billington argues that engineering art is different from architectural art. However, it is only if the aesthetics of architectural art has to have freedom from utility and resilience that the two really diverge. Previously, I have suggested four criteria to assess whether a bridge qualifies as public art. First, when I first see the bridge, do I experience a powerful emotional reaction? Second, is the bridge in total harmony with its context? Third, does the bridge give me a sense of frozen movement? Finally, does the bridge have a clear form that makes sense? Through these kinds of criteria, structural art and architectural art, aesthetics, utility, and resilience can be harmonized into public art.

There are three key points to take from this chapter. First, some kind of cultural disassociation happened during the Enlightenment, which continued to develop up until our final period of engineering history—art and science seemed to go their separate ways. This was highlighted by C. P. Snow in 1959 by his 'Two Cultures' Rede lecture, and it is still the topic of much debate. Second, a consequence of this disassociation has been that over the two centuries, engineering education has become narrower, more numerate, and more pragmatic; and structural engineering education has lost the quality of the aesthetic. Architectural education too often contains too little engineering. Third, architects and engineers can never be professionally disassociated. They must and do work together to build buildings. But sometimes they work in a state of tension born of mutual incomprehension.

Chapter 4

Understanding structure

Most of us, I suspect, have tried our hands at structural engineering by building a tower of children's play bricks. If you have then you will know that there is a limit to the height you can achieve. Intuitively you realize when you are getting close. At low heights a slight disturbance, say by just touching the tower, has little effect—it is quite stable. As the tower gets taller it gets more and more wobbly until it eventually topples. The height limit depends on a number of factors. For example the surface you are building on makes a difference—you can build a tower taller on a hard flat surface than on a soft fluffy carpet. In other words the foundations have to be suitable. Another factor is the size and shape of the bricks and, in particular, their cross section—the bigger the cross section the taller the tower. Yet another is whether each layer of bricks simply sits on the one below or is connected to it like a Lego brick.

In this chapter we will build on these familiar experiences to understand better how structures work. You'll remember that in Chapter 2 we showed (in Figure 9) the possible six degrees of freedom or independent directions (in line or rotating) in which a structure or any part of a structure can move or deform. We said that the key to structural art was to create pathways for forces to flow logically, and that the pathways are defined using these local or global directions when free movement is prevented. The restraint may be external, such as a foundation; or internal, such as a connection to another part of the structure. We found that a rope hanging in a catenary has only one degree of freedom and hence one force pathway, because it can only stretch as it resists tension. It cannot resist compression and can only resist a lateral force by changing its shape until it resists that force through direct tension (Figure 4).

Our toy brick tower also only has one degree of freedom and hence one force pathway—but now the force is compressive and the behaviour is more complicated. If we push down onto the top of a very low tower (say only one or two bricks) the bricks will simply squash under the load. You may be able to detect the amount of squash if you have soft foam or rubber bricks but you won't notice the very small squashing of wooden bricks—you would need a special instrument to measure it. Eventually, under a very heavy weight, the bricks will either squash flat or crush into fragments.

If we push down on a much taller tower then the behaviour is different. A tall tower will topple sideways rather suddenly. The bricks just want to get out from under the downward force or load so they take the path of least resistance and move sideways. As the towers get taller the load at which this happens will get smaller, until they reach the height limit and the tower falls without any load at all. Also the higher the tower the more sensitive it becomes to any slight sideways push or any tendency for the downward load being off-centre.

We can understand this better perhaps through another thought experiment—this time we'll push down at an angle so there is a downward and a sideways force. Imagine gently pushing the top brick of a tower sideways. Four things can happen depending on several different factors. First, if the bricks are smooth and the downward push load is small compared to the sideways push load, then the top one might just slide over the one underneath until its centre of gravity is just past the edge—and it falls. Second, if the contact surfaces between the bricks are rough then the friction between them will resist the sideways push—F, in Figure 11a. The net result is that a reaction force R_1 is generated that is a combination of the weight of the brick and the downward push (together labelled W) and the sideways force F. Furthermore, we can calculate R_1 using the triangle of forces that we introduced in Chapter 2—see Figure 11b. However it is important to realize that R_1 is only developed because the brick underneath stands firm. We can check that it does stand firm by going through exactly the same logic on the second brick down in Figure 11c. We can check the size of its reaction force R_2 using the triangle of forces but using $2W$ for the two bricks, as in Figure 11d. Then we can continue this process one by one for all of the bricks underneath until we get to the last one (in Figure 11e this is the 4th one) that sits on the ground with a reaction R_4 calculated in Figure 11f. If the ground can resist this reaction required from the bottom brick then the tower will stand.

A third possible behaviour of the tower is that the friction between the bricks is so great that the top brick doesn't slide but eventually starts to rotate about an edge, as shown in Figure 12a. Then as soon as its centre of gravity is poised just over the edge of the lower brick it will topple over, as shown in Figure 12b. Alternatively, if the bond between the bricks is strong enough then the whole tower might rotate about the bottom edge, as shown in Figure 12c. The degree of rotation (and hence the sideways force required) is much smaller for a taller rigid wall, as you can see from the diagram. A fourth possible group of behaviours for the tower is shown in Figure 12d, where the very bottom brick is prevented from rotating and the tower may fall, depending on whether the top one, two, three, four, or more bricks rotate together.

Now let's think about a quite different kind of tower—like a lamp post, a mast of a ship, or, more simply, a slender wooden cane

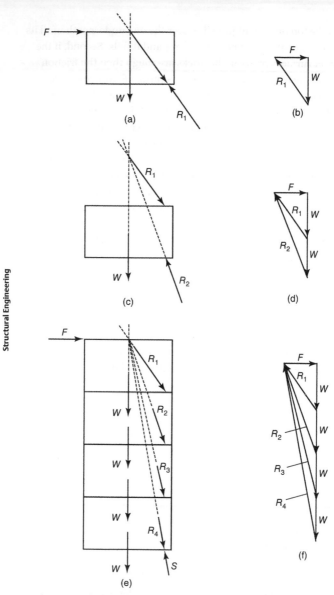

11a–f. Tower of blocks

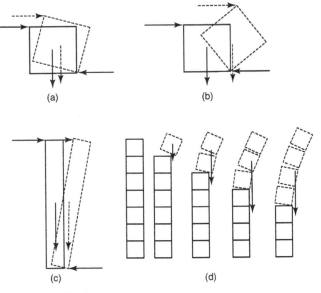

12a–d. Tipping blocks

pushed into the ground. Here we don't have separate bricks but just one continuous piece of wood which doesn't just sit on the ground but is firmly held by it, as in Figure 13a. Indeed, we will assume for the moment that the cane can't move or rotate at the bottom but, of course, the rest of the cane above the ground can deflect and bend. It's rather like a diving springboard in a swimming pool, except there the board is horizontal—other much more complex but real examples are an aeroplane wing and a tall building or skyscraper, which we look at in a bit more detail in Chapter 5. It is quite hard to visualize what is going on in all three dimensions so we'll simplify a bit. Figure 13a shows the cane bending due to a pushing force F only in the x direction—there are no forces in the y or z directions. The small element of the cane shown in Figure 13b has moved laterally in the x direction and rotated about the y axis, perpendicular to the paper. The turning force created by F (i.e. the moment that bends the element) will be F multiplied by the vertical distance from the horizontal line of action of *F* to the element. Consequently the maximum turning force or bending moment in the cane caused by *F*, will be at the base of the cane and equal to the force *F* multiplied by the total height of the cane. The ground must be able to resist that moment or the cane will rotate into the ground.

Unfortunately, real wooden cane towers don't just have one degree of freedom like our brick tower—they have all of the six we introduced in Chapter 2, in Figures 9 and 10. The forces and movements in the first three directions are perhaps the easiest to imagine. The first is simply pushing or pulling, a side to side movement in the *x* direction creating a tension or a compression in the *x* direction, as in Figure 10a. Then we may have a back to front movement in the *y* direction and an axial, up (pulling) and down (pushing or squashing) of the cane along its length in the *z* direction.

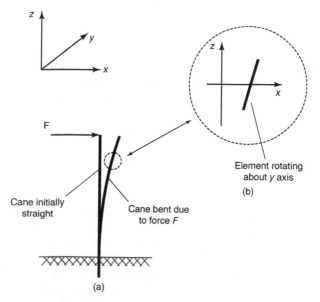

13a–b. Vertical cantilever

The movements in the second set of three directions are more difficult to visualize. They are the action forces on each face of the element of the cane that cause that element to rotate. These action forces are resisted by the reaction forces on the opposite face of the element and there is a pair for each axis. Together they cause the element to bend or twist as shown in Figures 10d–h.

We have now identified all of the forces acting along our six degrees of freedom on any piece of real structure we care to isolate. Whilst the elements in Figures 9 and 10 are rectangular blocks, in general our pieces of structure can be any size or shape, as seen in Figure 2. They can be anything from the small element of cane we have been examining so far, to any large piece of structure like the deck of a large bridge or the hull of a ship.

Some structures are relatively easy to determine

The next question is 'How do we find out how big the forces and movements are?' It turns out that there is a whole class of structures where this is reasonably straightforward and these are the structures covered in elementary textbooks. Engineers call them 'statically determinate'—we began to look at them in Chapter 2. For these structures we can find the sizes of the forces just by balancing the internal and external forces to establish equilibrium. In other words equilibrium between external and internal forces is both necessary and sufficient to determine the magnitude of them all.

Let's investigate this further by turning the cane on its side so we now have a cantilever beam. The cane is now like the diving springboard that we mentioned earlier. Imagine that F now represents a diver standing on the end of the horizontal springboard. We can find all of the internal forces in the springboard just by establishing equilibrium, because the cane tower and the springboard beam are statically determinate. In Figure 14a we have exposed the internal forces in the beam by

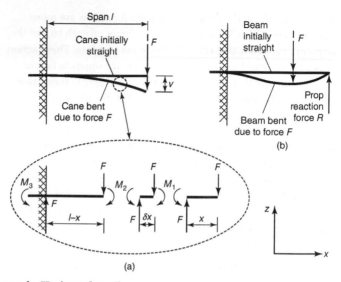

14a-b. Horizontal cantilever

yet another thought experiment. This time we mentally cut out an element of the beam and show the internal forces that balance the external force of F. The beam is of span length l and the element of length is δx. The element can be positioned anywhere along the length of the beam. In Figure 14a it is shown as a distance x from F. Let's now focus on the piece of beam (length x) between F and the first cut. We can see that there has to be a vertically upwards shear force to balance the weight of the diver F. At the cut there must be an equal and opposite shear force reaction on the right hand end of element δx. This is also true of the force downwards on the right hand end of the stub of beam $(l - x)$ connected to the support. Finally the support must also be able to supply a reaction F.

In just the same way, at the first cut there has to be a turning force or bending moment M_1. This occurs because the forces F on the outer piece of beam (length x) create a couple of F times x. Without M_1 this piece of beam would rotate freely. With M_1 in place it will still rotate but the amount of rotation is constrained because of the way it is connected to the rest of the beam. The equal and opposite moment on the left hand cut is M_2, bigger than M_1 because the lever arm of the couple has increased from x to $(x + \delta x)$ and so $M_2 = F \times (x + \delta x)$. At the fixed left hand support the bending moment is M_3 and equal to F times the whole span length of l so $M_3 = F \times l$.

Unfortunately many real structures can't be fully explained in this way—they are 'statically indeterminate'. This is because whilst establishing equilibrium between internal and external forces is necessary it is not sufficient for finding all of the internal forces.

To see why let's change the problem yet again. This time we decide to prop up the end of the cantilever as shown in Figure 14b. An example is a balcony with a column supporting the outer edge. A person stands somewhere near the middle of the span. Unfortunately this seemingly small change makes finding the internal forces much more difficult. We cannot now calculate the internal forces in the beam just by establishing equilibrium. You will recall that in Chapter 2 we said that Gaudí focused on equilibrium and did not consider the two other requirements of modern structural analysis—the material properties (or so-called constitutive equations that are part of the subject of 'strength of materials') and the geometry of deformation (the so-called 'compatibility equations'). These second two factors become important when structures are 'statically indeterminate'.

Let's think about this further using yet another example—this time the difference between a three-legged and a four-legged stool. When you sit on a three-legged stool, what are the forces in each leg? The stool is statically determinate, and the external and internal forces must be in equilibrium. The sum of the forces in each leg is equal to your weight. If everything is symmetrical and you are sitting centrally then each leg must take one-third of your weight, whether the floor is flat or not, and whatever the length of each leg. Of course if you are not sitting centrally then one of the legs may take a higher force than the others—indeed, if you were to sit exactly over one of the legs then it would have to take your whole weight (but this would be difficult to do in practice since you would have to balance yourself by constantly adjusting your position).

Now let's contrast our three-legged stool with a four-legged one. If you are again sitting centrally does each leg take one-quarter of your weight? The answer is, no. The four-legged stool is statically indeterminate. You will begin to understand this if you have ever sat at a four-legged wobbly table, say in a restaurant, which has one leg shorter than the other three legs. There can be no force in that leg because there is no reaction from the ground. What is more, the opposite leg will have no internal force either because otherwise there would be a net turning moment about the line joining the other two legs. Thus the table is balanced on two legs—which is why it wobbles back and forth. If you, like most people, insert some packing to stabilize the table then each leg will have an internal force. Alternatively, if the floor is not too uneven or stepped, you can often find a stable position by turning the table around its central vertical axis. It is then more difficult to find the magnitude of each force because each leg has one degree of freedom but we have only three ways of balancing them in the (x, y, z) directions. In mathematical terms, we have four unknown variables (the internal forces) but only three equations (balancing equilibrium in three directions). It follows that there isn't just one set of forces in equilibrium—indeed, there are many such sets. However, only one set will satisfy the other two requirements we mentioned earlier—the constitutive equations and the compatibility conditions.

Going back to our propped cantilever in Figure 14, we can now see why it is statically indeterminate. First of all, note that the local and global axes are the same and that there are no horizontal forces and hence no axial forces in the beam. Consequently there are only two degrees of freedom or only two ways of transmitting force within the beam—bending and shear. However, simply balancing the shear forces and the bending moments isn't enough because there are three unknown constraining forces—the prop reaction, the shear force, and the bending moment at the fixed end. So we have more unknown variables than we have equilibrium equations.

Another example is shown in Figure 15a, a two span beam resting on two supports with a person standing in the middle of each span. Many highway bridges are two span beams. The beam is statically indeterminate because we have (again, no horizontal loads) two degrees of freedom (moment and shear) and three unknown reactions R_L, R_C, and R_R. We can easily see that equilibrium is not enough by finding a set of forces in equilibrium that have incompatible displacements. For example in the middle diagram we obtain a set of forces in equilibrium just by thinking of each span as a simple beam with reactions at each end of $W/2$. Thus the total reactions are $R_L = W/2$, $R_C = W/2 + W/2 = W$ and $R_R = W/2$ as shown. So we have established equilibrium but, as the diagram shows, the corresponding displaced shapes can only happen if the beam cracks over the central support. If we enforce the condition that the beam doesn't crack, then the slopes either side of the central prop will be compatible but the reaction forces and internal forces in the beam will change.

Another example of statical indeterminacy is shown in Figure 15b. This is a pin-jointed square truss with one diagonal, no external loads, and a second diagonal that is too short to fit by an amount u. The structure without the ill-fitting diagonal is statically determinate. However, if we stretch the second diagonal to make it fit then internal forces are set up in the whole truss. An unknown force R is created in the second diagonal as we stretch it. We can use the triangle of forces (Chapter 2) to calculate all of the other internal self-balancing tensions and compressions in terms of R. But we cannot easily calculate the magnitude of R, since the

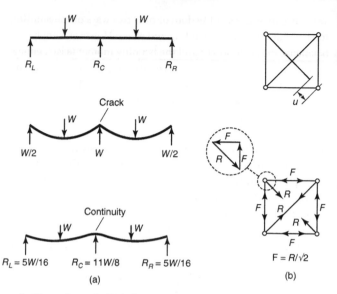

$R_L = 5W/16$ $R_C = 11W/8$ $R_R = 5W/16$

(a)

$F = R/\sqrt{2}$

(b)

15a–b. Forces in statically indeterminate structures

stretched second diagonal and the rest of the truss will deform until all of the forces are in equilibrium and all of the deformations (stretching of tension members and shortening of members in compression) are compatible.

Determining the indeterminate

Detailed analyses of these statically indeterminate problems are beyond the scope of this book, but we will look at some of the basic principles. As we have said, we have to apply the three conditions: equilibrium, constitutive relations, and compatibility of displacements.

To understand this, we need now to look at equilibrium slightly differently. You will recall from Chapter 2 that when a structure is in equilibrium it has minimum strain energy. First, it is important to note that this has nothing to do with the first law of thermodynamics and the principle of the conservation of energy. These laws require that the total energy is constant but do not refer to the way it changes. By contrast the various theorems that are used to analyse statically indeterminate structures refer not to total energy but only to changes in energy. Specifically they refer to maximum and minimum values. One way of thinking about this is to imagine an undulating surface, like a mountain range, with peaks and troughs. The surface represents our structure under various forces and displacements. The height of the surface represents the energy at a point on the surface, and the peaks and troughs represent points of equilibrium—points where the energy is a local maximum or minimum. Now imagine a ball rolling on this surface and perhaps balancing on a peak or settling in a trough. Wherever the ball is balanced we are going to change its position by perturbing it slightly. If it is balanced on an energy peak, rather like a ball balanced on a seal's nose, then it will roll away from its equilibrium position (the seal can only maintain balance by moving its nose). This is called 'unstable equilibrium'. It is the sort of equilibrium you experience when you stand on one leg or balance on two legs of a three-legged stool. We will look at it further in Chapter 5. For now imagine that the ball has settled in one of the troughs—perturb it and it rolls back. This is called stable equilibrium and is the most desirable form. However, in both cases, stable or unstable, the equilibrium is local minimum or maximum because a very large movement or perturbation might push the ball over a local peak and into a new and different part of the energy surface.

There is a further complication too. Strictly speaking, minimum strain energy as a criterion for equilibrium is true only in specific circumstances. To understand this we need to look at the constitutive relations between forces and deformations or displacements. Strain energy is stored potential energy and that energy is the capacity to do work. The strain energy in a body is there because work has been done on it—a force moved through a distance. Hence in order to know the energy we must know how much displacement is caused by a given force. This is called a 'constitutive relation' and has the form 'force equals a constitutive factor times a displacement'. The most common of these relationships is called 'linear elastic' where the force equals a simple numerical factor—called the stiffness—times the displacement (see Figure 16a where the graph is a straight line). A material is elastic if, on decreasing the force (unloading), the graph follows the same path back. The inverse of the stiffness is called flexibility and if the graph is a curve then the elasticity is

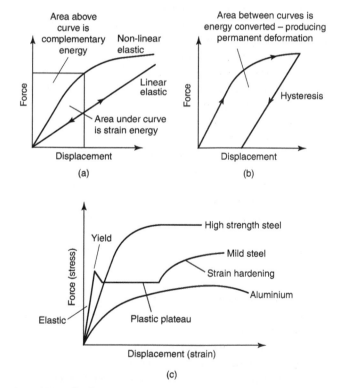

16a–c. Force displacement relations

68

non-linear. The strain energy can be calculated for a given force because it is simply the area under the graph as shown. The area above the curve also turns out to be important in structural analysis and is called the complementary energy. For a linear elastic material the strain energy equals the complementary energy because the areas are the same. A last complication is that many real materials do not follow such simple curves (see Figure 16c) or their unloading path is a loop as Figure 16b—a phenomenon called hysteresis. In this case some of the energy is converted into other forms such as heat and sound (forms of kinetic energy) as permanent deformations are created.

Solutions using virtual work

There are two approaches to satisfying the three conditions, and both are very general and very powerful although they only apply to elastic materials as Figure 16a and not as in Figure 16b. They again depend on thought experiments. The first approach relies on imaginary or *virtual displacements*. We impose these virtual displacements on our structure. In other words we imagine every node moving an amount dictated by the chosen set of virtual displacements. We then see what happens when we vary them whilst keeping the forces unchanged. We do this to find the equilibrium state using a principle we'll come to in a moment. The virtual displacements are chosen so that the external and internal displacements are compatible but they may result in forces that are not in equilibrium.

For the second approach we choose to apply imaginary or *virtual forces*. We impose these virtual forces on our structure. We then see what happens when we vary them whilst keeping the displacements unchanged. We do this to find the equilibrium state, again using a principle we'll come to in a moment. The virtual forces are chosen so that the external and internal forces are in equilibrium but they may result in incompatible displacements, as we saw earlier for the two span beam of Figure 15a.

The two principles we use to find a state of equilibrium when we vary the virtual displacements or virtual forces are: first, 'the principle of the minimum total potential'; and, second, 'the principle of the minimum of total complementary potential'. As these statements imply, when the total potential or the total complementary potential is a minimum our structure is in equilibrium. Total potential consists of the internal strain energy of the structure plus the potential energy of the external action forces (loads) on the structure. Complementary potential is the internal complementary energy of the structure plus the potential energy of the external reactions. (In passing, note that the latter is often zero if the support reactions do not move).

We can use either of these principles with any arbitrary set of action and reaction *forces* that are in equilibrium—we'll call that set F. We also use a quite separate, arbitrary, and independent set of external *displacements* of the loads and supports that are compatible with the internal displacements of the members—we'll call that D. It is important to note that F and D are totally separate and independent—so we are at liberty to choose them both to our advantage in any particular calculation.

In a proof beyond the scope of this book it can be shown that for any F with any D the virtual work done by the external loads and supports will equal the virtual work done by the internal members as they strain. We call the first the external virtual work and the second the internal virtual work. The only condition on the proof of these theorems is that the displacements are small when compared to the dimensions of the actual structure.

How do we use the first principle of minimizing the total potential? We can choose F to be all of the *known* actual external forces on the structure. Then we can choose D to be any set of quite arbitrary (but convenient for our purpose) set of virtual displacements. We then calculate the strain energy in all of the internal members of the structure as they undergo the imposed virtual displacements. We calculate the potential of the external loads as the virtual work done by those loads in moving through the virtual displacements. We add the strain energy and the potential of the external loads to get the total potential. Then, as we noted earlier, we keep the forces constant and vary the virtual displacements so that, by using some appropriate mathematics, we can find the set of virtual displacements that minimize the total potential. At that stage the problem is solved because the external forces are the actual forces, the virtual displacements are compatible, we have used the constitutive relationships (in finding the total strain energy) and the structure is in equilibrium. The calculated virtual displacements are therefore the real displacements.

The second principle is minimizing the total complementary potential. Now we choose D to be the actual displacements of the structure. Then we choose F to be a set of quite arbitrary virtual forces and proceed in a similar manner. In the calculation the displacements are kept constant and the virtual forces are varied. Eventually using appropriate mathematics we can find a set of virtual forces that minimize the total complementary potential. So the problem is solved because the displacements are the actual displacements, the virtual external and internal forces are in equilibrium, we have used the constitutive relationships (in finding the total strain energy) and the total structure is in equilibrium. The calculated virtual forces are therefore one possible set of real forces.

The finite element method

Almost all modern structural calculations are performed on a computer. The most well known tool is 'the finite element (FE) method'. To get a flavour of it here, we will consider only one version, which assumes linear elastic materials and minimizes the total potential. It is commonly called 'the stiffness method'.

In brief, the total structure is divided into discrete finite elements or pieces of structure, as we have discussed before. The elements or pieces may be any size but are usually either a whole structural member, such as a beam, or imaginary divisions (of a plate, wall, or floor) in suitable shapes such as triangles and rectangles. The elements are connected only at joints or nodes. These are normally at the ends of a member or the corners of the imaginary triangles or rectangles. External forces are applied at the nodes only. Displacements within each element are calculated using an assumed mathematical relationship with the displacements at the nodes and continuity between adjacent elements. That relationship is called a mathematical shape or displacement function.

Constitutive equations are derived for each separate internal force in turn, along each separate local degree of freedom of the element, at a node. They are called 'influence stiffness coefficients' because they are calculated by isolating each node of each member. A virtual displacement is imposed along a degree of freedom at that node with all other forces kept constant. By equating the consequent internal and external virtual work a relationship is developed between a force and a corresponding displacement in a local pathway or degree of freedom. These separate relationships, of force equals a stiffness factor times a displacement, can then be added or assembled together in a large set of simultaneous equations. There is one equation for each global degree of freedom. We can think of each equation as representing the relationship between the external force along one global pathway at a node and the corresponding unknown displacement such that the known force equals the known stiffness times the unknown displacements at that node.

We humans find solving large sets of simultaneous equations very difficult. Computers do it easily. Once we have the unknown displacements the computer uses them to calculate all of the internal forces in the structure.

It is important to note that all of this can only be done because the matrix equations are linear—they involve no variables that are squared or cubed. We are relying on the principle of superposition—so called because the various parts can be superimposed together. This is not true for non-linear systems, so extended techniques have to be used. For example we can apply loads in small increments, then reset and solve the equations at each step.

There are three key points to take from this chapter. First, some simple structures (as in some of the more elementary textbooks) are statically determinate, so we can calculate the internal forces by just making sure all of the forces are in equilibrium. Unfortunately, more complex structures have too many unknowns to be solved this way—they are statically indeterminate. We find the internal forces in these structures by satisfying the three conditions of equilibrium, constitutive relations, and compatibility of displacements when the total potential is a minimum.

In the next chapter we'll explore how these ideas are used to design and build very large structures such as jumbo jets and cruise liners.

Chapter 5
Movers and shakers

What have a jumbo jet, a cruise ship, a long span bridge, and a tall skyscraper got in common? Answer: they are all gigantic tubes. Their similarities and differences will reveal much about their main functions, the forces they must resist, and the ways in which they have to be strong.

In previous chapters we have assumed that structures do not move. In this chapter we will explore some big structures that do move, and some complex phenomena including major forms of instability such as buckling, resonance, and flutter.

The form of a big structure depends heavily on the basic job it is required to do and the forces it has to resist. Ships must float, aeroplanes must fly, bridges must carry traffic, and buildings must provide shelter.

Archimedes' Principle explains why ships float. The weight of any heavy object in water is countered by a reaction which is the upward force of buoyancy. If the object weighs more than the water it can displace, it will sink. However, sea water is seldom still and so the buoyancy forces vary with the waves (Figure 17a). For example if the sea wavelength just happens to be the same length as the ship then the ship is effectively spanning between the wave peaks rather like a simply supported bridge beam. Engineers and naval architects commonly call that sagging. Alternatively, if the wave peak is at the centre of the ship then much of the weight is reacted at the centre, with fore and aft portions of the ship acting as cantilevers—that is called hogging. The other main forces on a ship are the forward thrust generated by its propellers and drag from the resistance of the water and wind.

Giant cruise ships can appear rather top heavy (Figures 17b–c). To understand how they are stable, we need to define something called the centre of buoyancy—the centre of gravity of the water displaced by the ship. A key attribute for any ship is the position of the centre of gravity of its weight relative to its centre of buoyancy. Much of the weight of a cruise ship is in the hull. The vertical weight and buoyancy forces are equal, and the ship floating in still

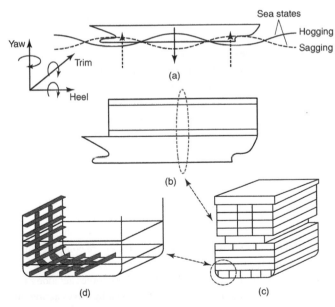

17a–d. Ship structure

water will take up a position such that the centres of weight and buoyancy lie on the same vertical line. When the ship rolls or heels, the centre of buoyancy moves sideways so the vertical weight down and the vertical buoyancy force up are no longer in line. This misalignment creates a turning force—a moment. In effect the centre of buoyancy rotates about a point called the metacentre. If the metacentre is above the centre of gravity of the weight of the ship, then the ship is in stable equilibrium. This is because the turning moment resists the roll and the ship will right itself. If the metacentre is below the centre of gravity as the heel angle increases, the ship is in unstable equilibrium and will become top-heavy and capsize—something to be avoided even when the hull is badly damaged.

Perhaps one of the wonders of modern engineering is that massive structures such as jumbo jets can fly. One way of understanding how planes fly is to perform our own experiment. Take a piece of A4 paper and hold it lightly and horizontally on the long side but then tilt it very slightly to a small upward angle—engineers call it the angle of attack. Now blow on the short edge of the paper and watch it lift. The reason it lifts is that as air passes over the paper (the wing) the pressure reduces on the top and this, coupled with the air pressure underneath, provides the lift force. Of course in an aircraft the wing moves through the air whereas in our experiment the air moves over the paper—but the effect is the same.

Aeroplanes take off or ascend because the lift forces due to the forward motion of the plane exceed the weight—see Figure 18a. In level flight or cruise the plane is neutrally buoyant and flies at a steady altitude. The distributions of the forces, together with thrust and drag, over the surface of the aircraft are complex but can, like ships in an overall sense, be represented by the lines of action shown in Figure 18a. Rarely do these forces act along the same line so the moments have to be balanced by forces acting on the tail plane.

The structure of an aircraft consists of four sets of tubes: the fuselage, the wings, the tail, and the fin. For obvious reasons their weight needs to be as small as possible. Early aircraft were truss structures of wood or steel covered for example with fabric, which was not part of the structure. Modern aircraft structures are semi-monocoque—meaning stressed skin but with a supporting frame. In other words the skin covering, which may be only a few millimetres thick, becomes part of the structure. For example one of the critical loading conditions during flight is the difference between the air pressure in the cabin and the much lower pressure outside, and this is resisted by the stressed skin much like a toy balloon resists the air pressure inside when you blow it up. In Figure 18b and c the skin has been omitted to show, in diagrammatic form, the underlying closely spaced frames and stringers, but we should remember that these act structurally together with the skin.

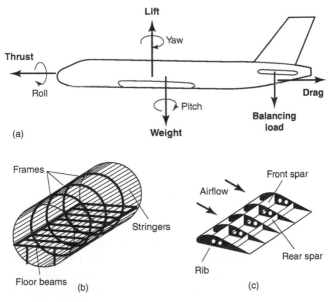

18a–c. **Aircraft structure**

In an overall sense, the lift and drag forces effectively act on the wings through centres of pressure. The wings also carry the weight of engines and fuel. During a typical flight, the positions of these centres of force vary along the wing—for example as fuel is used. The wings are balanced cantilevers fixed to the fuselage. Longer wings (compared to their width) produce greater lift but are necessarily heavier—so a compromise is required. Figure 18c shows a typical stressed skin wing structure in which the ribs have holes not just to reduce weight but also to allow the flow of fuel between compartments.

Figure 19a–b illustrates a large suspension bridge such as the innovative first Severn Bridge built in the UK in 1966. It was one of the first modern suspension bridges to use a long tube as the bridge deck—a box girder. Box girders have to resist loads similar to those on a ship or aeroplane—in particular, weight and uplift forces. Uplift on a suspension bridge girder is provided by the internal tension forces in the vertical suspender cables (which are actually inclined in the Severn Bridge) that hang from the main curved suspension cable. The main cable is rather like a very large washing line suspended from two props—the bridge towers.

Clearly, unlike ships and aeroplanes, a bridge does not have any thrust forces from engines. Indeed box girders are the very inverse of aeroplane wings because we want them to remain still and not to lift. However, as the wind blows, the girder will tend to sway from side to side. Modern bridge box girders are tested in wind tunnels to check that the differential pressure on aeroplane wings is absent.

The final example in our quartet of large structures is a skyscraper tower block. Again there are many forms but all are vertical cantilever tubes or boxes. Some consist of a central stiff core tube structure usually of concrete and containing elevators and stairs. The floors and outer walls are then built out from the core. Other skyscraper tubes have an outer stiff wall or rim of closely spaced

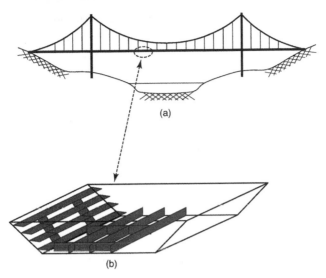

19a–b. Bridge structure

columns and are called framed or trussed tubes. Perhaps the most famous example was the New York World Trade Center brought down by terrorists on 11 September 2001. More recent structures are bundles of tubes tied together such as the Burj Khalifa opened in 2010.

Uplift forces for skyscrapers come from the foundations and the ground below. Ground engineering and the geology of the sub-soil are crucial, and structural engineers often rely on specialist geotechnical engineers. Apart from the self-weight of the building, the two main forces are due to the lateral load from the wind and possible shaking from the ground from earthquakes in seismically active zones such as California. Such buildings can no longer be treated as static objects.

The loads or demands on any large structure during its life cycle are complex. As we will see in a little more detail in Chapter 6, engineers have to estimate the extremes of demand to ensure safety, and these have in the past largely been set by regulatory rules. For example aeronautical engineers use a 'manoeuvring envelope' to show particular combinations of speed and load during take-off, cruising, adjusting position in flight, and landing. Good practice is set out by airworthiness authorities such as the European Aviation Safety Agency, classification societies such as Lloyds Register for ships, and national and international codes of practice for bridge and building structural engineers. In more recent years there has been a trend towards performance-based standards where safety targets are prescribed leaving engineers free to choose how to meet them.

Dynamic equilibrium

All four of our structures have to be designed for dynamic movement. As you will recall from Chapters 2 and 4, static equilibrium occurs when the forces on a body balance each other—the body is static. However, static equilibrium is actually the end point of a process. It is a process in which external loads are applied steadily and slowly. The changes in each piece of a static structure as loads are applied are of two kinds. The first is slow movement as a whole rigid body. The second is a set of slow internal changes (lengthening, shortening, shearing, twisting, and bending) that create the internal forces we have been discussing.

When structures move quickly, in particular if they accelerate or decelerate, we have to consider two extra forces and some new phenomena. In brief, the two new forces are the inertia force and the damping force. They occur, for example, as an aeroplane takes off and picks up speed. They occur in bridges and buildings that oscillate in the wind. As these structures move the various bits of the structure remain attached—perhaps vibrating in very complex patterns, but they remain joined together in a state of dynamic equilibrium.

An inertia force results from an acceleration or deceleration of an object and is directly proportional to the weight of that object. You can appreciate it by thinking of what happens when you travel in a car. When you are cruising at a uniform velocity both you and your car have momentum but you feel no inertia force. If you accelerate your car you will feel an inertia force—it pushes you back into your seat. Likewise, if you have to stop very suddenly you will be thrown forwards. All parts of your car will experience inertia forces too. Newton's 2nd Law tells us that the magnitudes of these forces are proportional to the rates of change of momentum. Since momentum is mass times velocity, then any inertia force on any part of the car, or your body, is directly proportional to the mass of that part and will be transmitted through the structure along the pathways of the degrees of freedom.

The second new force is damping—a resistance to motion. Damping arises from friction or 'looseness' between components. As a consequence, energy is dissipated into other forms such as heat and sound, and the vibrations get smaller. You will recall that the strain energy in a structure is due to the change in position of the ends of a member. Moving bodies have kinetic energy, which depends on momentum—indeed it depends on the mass and the square of the velocity. The kinetic energy of a structure in static equilibrium is zero, but as the structure moves its potential energy is converted into kinetic energy. This is because the total energy remains constant by the principle of the conservation of energy (the first law of thermodynamics). The changing forces and displacements along the degree of freedom pathways travel as a wave—rather like sound travelling through water or air, or light and radio waves through space. The amplitude of the wave depends on the nature of the material and the connections between components.

Examples of damping are the shock absorbers of your car designed to give you a smoother ride. You may recall that the London Millennium Bridge was closed soon after it was opened in

2000 because of excessive vibrations. To combat this, the design engineers installed mechanical dampers, which absorb and dissipate energy, to control them.

As discussed in Chapter 4, there are three conditions for the integrity of a structure—equilibrium, constitutive relations, and compatibility of displacements—and we need to use theories that rely on internal and external virtual work and strain energy to find equilibrium conditions for statically indeterminate structures. For moving structures, we have to modify the equilibrium requirement from static equilibrium to dynamic equilibrium—in other words the flow of forces along each pathway of a degree of freedom has to contain the two extra forces of inertia and damping. Thereafter, thanks to a theorem attributed to French theorist Jean-Baptiste le Rond D'Alembert (1717–83), we can use the same virtual work and potential analyses as we did for statics.

You may wonder how a moving object can be in equilibrium. There are two answers. First, as Einstein demonstrated in his theory of relativity, motion is relative and so depends upon a reference system. All parts of the moving body are at rest with respect to each other—they are moving together. Second, D'Alembert's Principle applies virtual work to a set of forces (including inertia and damping) not to the state of motion of a body. Displacements are virtual and not real, and are applied momentarily at a particular instant of time.

Looking down the structural hierarchy

We started the chapter by saying that at the very top level of our structural hierarchy all four structures, the ship, the aeroplane, the bridge girder, and the skyscraper, are tubes or boxes that act as beams. Each beam must be capable of resisting the overall forces we have identified. So ships are beams—hogging or sagging as in Figure 17a and twisting. A parked aeroplane will be a beam supported on its wheels—during flight the wings are cantilever beams. A bridge deck girder will bend as a multi-span beam and a skyscraper is a vertical cantilever beam.

At lower levels of the hierarchy we must look in more detail at how the various pieces or elements of the structure will behave. For example a cruise ship is a set of multi-layered tube or box beams. The lowest beam is the hull, which contains a number of lower decks. On top of the hull sits another set of much lighter beams which are each of the decks of the superstructure as shown diagrammatically in Figure 17c. The top of the box is a deck floor and bottom of one beam and the roof of the beam below. The sides of the beams are plates stiffened as shown in Figure 17d.

As we have already noted, an aeroplane is basically four beams, the fuselage, the wings, the tail, and the fin. The fuselage beam is made up of frames spaced at intervals as shown in Figure 18b. The frames are connected by longitudinal members called stringers or longerons. They transfer the load acting on the outer skin onto the frames. The equivalent members in the wings are called spars and ribs as shown in Figure 18c.

The top plate of the bridge box girder carries the traffic   the tarmac and street furniture are laid directly on it. The stiffeners shown in Figure 19b on the sloping web and on the bottom plate are repeated all around the inside of the girder including the top deck (not shown). So if the bridge deck plate were transparent as you looked down into the girder, you would see a forest of stiffeners and regularly spaced diaphragms.

In Chapter 4 and Figure 14 we identified the forces in a simple cantilever beam acting in one plane only with just three degrees of freedom—movement in the x and z directions and rotation about the y axis. Now we need to look at what happens when a beam like this hogs or sags—we'll again keep it simple by thinking about only three degrees of freedom. When the beam hogs, the top fibres or elements of the material of the beam will stretch in tension and the bottom fibres will compress. As it sags the opposite is true. You can see this for yourself if you hold a length of rubber bar and bend it in your hands.

For the structure to be safe the materials must be strong enough to resist the tension, the compression, and the shear. The strength of materials in tension is reasonably straightforward. We just need to know the limiting forces the material can resist. This is usually specified as a set of stresses. A stress is a force divided by a cross sectional area and represents a localized force over a small area of the material. Typical limiting tensile stresses are called the yield stress (Figure 16c) and the rupture stress—so we just need to know their numerical values from tests. Yield occurs when the material cannot regain its original state, and permanent displacements or strains occur. Rupture is when the material breaks or fractures. You may recall that in Chapter 1 we said that paper is strong in in-plane shear—the same is true for the other materials used in big structures. Limiting average shear stresses and maximum allowable stress are known for various materials.

There are several forms of instability

Strength in compression is much more difficult—we looked at it in Chapter 1 and again in Chapter 4. Buckling is the first major form of instability we are going to examine in this chapter—you can remind yourself how it happens with a long slender bar such as a ruler. The material of the bar simply tries to get out of the way of the squeezing by moving sideways along the path of least resistance. The bar is stable until a critical load is reached when it bifurcates (i.e. it takes one of two possible paths of least resistance) and its stiffness suddenly drops. If the bar is a plate then the path of least resistance is normal (at right angles to) to the plane of the plate, in other words in the direction of the plate thickness. This is a problem, particularly for aeroplane structures, because the thin skin will buckle easily. It is also an issue for ships, bridges, and skyscrapers because, although the skins are thicker, the breadth to thickness ratios of the plate panels—which determines their proneness to buckling—can be similar. Therefore, in every case the skin plates are not enough on their own—they must be stiffened as shown diagrammatically in Figures 17d, 18c, and 19b. In each diagram, for the sake of clarity, the skin has been omitted and only a partial set of plate stiffeners shown. In actual ships, aeroplanes, and bridge box girders the stiffening is more complex than shown in the diagrams (e.g. stiffeners may be plates (as shown), or tee-bars, angles, or bulb flats) but the principles are the same.

To understand how the stiffeners work you may recall that, as described in Chapter 1, if we fold a piece of paper (the skin plate) into a zig-zag shape it becomes much stiffer in bending and compression. The rectangular bar stiffeners serve the same purpose—they effectively increase the cross sectional area of the plate and its resistance to bending. They are joined to the plate, usually by welding, and so act with it as a structural element.

Another thought experiment helps us to see why. To do it we are going to drop down another level in our structural hierarchy. Instead of thinking about the fuselage or wing as a beam we now need to think about a piece of stiffened plate that is an important part of the beam. Mentally we will cut out a length of one stiffener with some plate attached to it on either side. This gives us a long bar with a cross section which is a tee shape. If the bar is taken from the bottom plate as shown in Figure 19b then it will be an inverted tee. If it is taken from the side panelling the tee will be on its side. If it is taken from the deck plates (not shown) the top of the tee is the deck plate—which engineers call the flange. The stiffener forms the vertical part of the tee and is called the web.

Now that we have isolated the tee bar we can examine its likely behaviour in compression or in bending. Engineers can find, by calculation and by laboratory testing, the total compression force that makes the tee bar buckle. They then make sure that this force is greater than the compression force that the bar experiences by being part of the wing or fuselage beam. Of course in the beam our stiffened plate section is not actually free along its length—it is attached to the rest of the tube. A tee bar may also twist. So our method is an approximation—it is a model to help us make a decision as to whether the tube will be safe—it is not the reality. Modern practice using the finite element method enables us to make theoretical estimates that do allow for compatibility between our tee bar and the rest of the structure but it is still approximate because of the simplifications necessary to do the computer analysis that we discussed in Chapter 4. One of the challenges to engineers who rely on finite element analysis is to make sure they understand the implications of the simplifications used.

A bar in axial compression usually deforms into the typical shapes shown in Figure 20a–e. The actual shape largely depends on the number of degrees of freedom restrained at the ends. For example in Figure 20a the bar has pinned ends and is commonly used as a base case for making comparisons. In Figure 20b the bar is fixed at one end and free at the other. As it buckles with a shape that is one-half of the pin-ended case, engineers say that it has an effective length of one-half. In Figure 20c the bar is fixed at both ends and has an effective length of 2 (because its shape is twice the length of the pin-ended case).

As the load is increased the bar eventually reaches a critical load at which it loses all stiffness and lateral displacements will grow large very quickly. 'Snap-through' buckling occurs when a bar jumps suddenly into another, usually symmetrically opposite, state.

Another kind of buckling can occur in beams with an I-shaped cross section. Beams with a deep vertical web of the I and smallish flanges top and bottom of the I may have low stiffness to twisting—engineers say that these beams have low torsional stiffness. A phenomenon called lateral torsional buckling can happen in such beams at

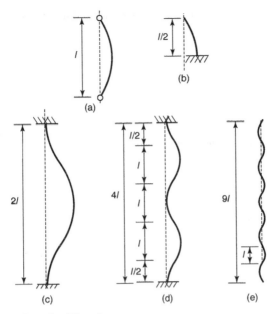

20a-e. Column buckling shapes

quite low levels of stress. The flange in compression moves sideways (lateral) because it is being squeezed, but the opposite tension flange does not. The beam cross section therefore twists (torsional) until the whole section loses stiffness and buckles.

So far we have not recognized that slender tubes can lose their cross sectional shape rather easily as they bend and twist. You will recall that at the end of Chapter 1 we saw how a long slender cardboard tube can warp. Each of our tubes or boxes maintains its shape because it is stiffened at regular intervals across the whole cross section by partitions, diaphragms, or bulkheads. Normally these bulkheads have holes within them to allow for the movement of people or material. Often they are fitted with doors to maintain water tightness—especially as a safety measure if the hull is breached.

One significant difficulty in a modern cruise ship is the stiffening of the hull plates around very large open spaces such as a theatre or atrium that may extend through two or three decks or more. Likewise the aeroplane fuselage where the wings are attached has also to be considerably stiffened to transmit the bending forces from the cantilever wings. The design around such large openings has to be considered very carefully.

Dynamic loads cause vibrations. One particularly dangerous form of vibration is called resonance—the second form of instability we are looking at in this chapter. All structures have a natural frequency of free vibration. You will have witnessed this if you have ever pushed a child's swing. You rapidly learn when to push for maximum effect because the swing goes back and forth at its natural frequency. If there is a lot of damping then the vibrations will reduce rapidly when you stop pushing. If the damping is low the swing will continue to carry on for some time.

Resonance occurs if the frequency of an external vibrating force coincides with the natural frequency of the structure. The consequence is a rapid build up of vibrations that can become seriously damaging. One of the most well known examples is soldiers marching across a bridge, when they must break step since if they do not the bridge might resonate. Forced vibration loads may arise from engine vibrations, waves slamming, wind gusts, buffeting, people walking and dancing, and even Mexican waves in sports ground stadia.

Wind is a major source of vibrations. As it flows around a bluff body the air breaks away from the surface and moves in a circular motion like a whirlpool or whirlwind as eddies or vortices. Under certain conditions these vortices may break away on alternate sides, and as they are shed from the body they create pressure differences that cause the body to oscillate. As we have noted in Chapter 4, a structure is in stable equilibrium when a small perturbation does not result in large displacements.

A structure in dynamic equilibrium may oscillate about a stable equilibrium position.

In all that we have considered so far the displacements of the structure have been very small compared to the dimensions of the structure. Sometimes this is not so. For example, if a piece of structure, such as a column, is so slender that its lateral displacement is significant, then an extra turning force or moment is created, because the line of action of the force and the position of the centre of the column are now misaligned. Engineers call it a P-delta effect. 'P' is the compression force and 'delta' is the newly significant lateral displacement. This can lead to premature instability in, for example, skyscraper buildings, if not allowed for.

Our third form of instability is related to this and is called flutter. Flutter is dynamic and a form of wind-excited self-reinforcing oscillation. It occurs, as in the P-delta effect, because of changes in geometry. Forces that are no longer in line because of large displacements tend to modify those displacements of the structure, and these, in turn, modify the forces, and so on. In this process the energy input during a cycle of vibration may be greater than that lost by damping and so the amplitude increases in each cycle until destruction. It is a positive feed-back mechanism that amplifies the initial deformations, causes non-linearity, material plasticity and decreased stiffness, and reduced natural frequency. The dramatic Tacoma Narrows bridge disaster of 1940 is one of the most famous examples when a wind of only 42 miles per hour (mph) caused large longitudinal and torsional oscillations that eventually brought down the bridge.

Regular pulsating loads, even very small ones, can cause other problems too through a phenomenon known as fatigue. The word is descriptive—under certain conditions the materials just get tired and crack. A normally ductile material like steel becomes brittle. Fatigue occurs under very small loads repeated many millions of times. All materials in all types of structures have a fatigue limit. If in each flight, cruise, or truck journey over a bridge the stress is below that limit, then fatigue damage will not occur. Above that limit, even though the stresses are still very small, under certain conditions, unseen cracks can start to form and grow a little on each cycle of load. Fatigue damage occurs deep in the material as microscopic bonds are broken. The problem is particularly acute in the heat affected zones of welded structures. This is because welds tend to have micro-cracks that can't be seen with the naked eye. The welding process causes metallurgical changes that can make the steel more brittle. Welding requires a lot of heat, and the consequent differential cooling of the metal can cause residual stresses also leading to problems. Residual stresses are complex self-balancing, locked-in internal stresses retained after a piece of material has been strained from either heat or external force.

Engineers look for a relationship between a stress level in a particular material like steel and the number of cycles to cause fatigue cracking. They test specimens of material or specific parts of a structure by loading them many millions of times—usually in a laboratory rig. They vary the stress levels and plot a graph of the number of cycles to failure for a given stress level. The graph is known as an S-N diagram, where S stands for the stress level and N the number of cycles to failure. Unfortunately, there is a lot of scatter in such graphs, making them difficult to use—nevertheless, they are often used to calculate an estimate of how long a bridge structure can last before there is significant fatigue cracking—its fatigue life.

Suddenly applied loads like a hammer striking a nail or a pile hammer driving a pile into the ground can exceed maximum slowly applied static buckling loads without inducing large bending strains or deformations. This is because the impact produces pulse wave forms such as shown in Figures 20d–e with wavelengths that depend on pulse amplitude.

There are three key points to take from this chapter. First, a jumbo jet, a cruise ship, a long span bridge, and a tall skyscraper are all, in essence, gigantic tubes or boxes that act as beams. Then as we look down the structural hierarchy we see the differences of detail. Second, structures can be in dynamic equilibrium, where connected structural parts move as one whole structure, but here we must consider two extra internal forces—inertia and damping. Third, there are several complex forms of instability including buckling, resonance, and flutter.

In the face of such complexities predicting everything that might happen—foreseeing the complete life cycle—is a challenge. In the final chapter we will look at the risks and benefits of the business of structural engineering and the need for systems thinking.

Chapter 6
Resilience

You can never say, 'never'. If something can happen, then there is a finite risk that it will. Structural engineering is safety critical. In other words if a structure fails then people are very probably going to be killed. Structural engineers face some central questions: What are the risks? How big are they? How safe is 'safe enough'?

As we said in Chapter 1, the safety of the structure of a ship, aeroplane, bridge, or skyscraper is necessary but not sufficient for the safety of the whole ship, aeroplane, bridge, or skyscraper. In other words if the structure fails then the whole system fails, but the whole system may fail for other reasons than those to do with structure—an aeroplane may crash because of pilot error, a ship may drift out of control through a loss of power, and a skyscraper may be gutted by fire.

Structural engineering was once defined, whimsically, as 'The art of moulding *materials* we do not really understand into *shapes* we cannot really *analyze*, so as to withstand *forces* we cannot really *assess*, in such a way that the *public* does not really suspect'. This tongue in cheek definition does include six of the essential ingredients (in italics) of good structural engineering and does make clear the uncertainties and risks. For example in previous chapters we have seen that *materials* have to be strong enough to cope with the flow of internal forces through the *shapes* of the structure. We have learned something of how engineers *analyse* the flow of *force* and consequent displacements using methods such as virtual work.

Now we need to focus on *assessment* and the role of the *public*—you and me. Risk, uncertainty, and safety are central issues. In Chapter 1, we pointed out that engineering is not a science, but engineers use science to design and build structures. Because all forms of knowledge, including science, are incomplete, judgements have to be made. Given these uncertainties how do engineers assess a degree of safety? For example should an estimate of the strength of a structure exceed an estimate of the biggest load by 10 per cent or 50 per cent or 100 per cent—or even more? Does it just depend on the uncertainties in our estimates of load and strength or are there other things to consider as well? The answers to these questions are important to us all—for the simple reason that we use the structures and we are the ones who may be killed if something goes wrong.

The questions go further than personal safety. In the 21st century, we humans face some difficult and ethical challenges—particularly about our responsibilities to our children and unborn future generations. Principal amongst them are the need to use resources sustainably, to minimize embodied energy (i.e. the sum of all the energy required to make something) to mitigate global warming, to manage the possible effects of climate change by making structures robust against extreme natural events like severe storms, and to protect people from extreme acts of terrorism. At the same time engineers must continue to work within the scope of financial and business needs, provide clients with what they need, ensure proper maintenance, and work in accordance with statutory regulations.

Resilience is the key

We began to appreciate in Chapter 2 that construction risk, safety and vulnerability, and use of energy are all aspects of one very important word—'resilience'. Resilience is the ability of a system to recover quickly from difficult conditions. Resilient people seem unflappable—they cope with, and recover from, severe setbacks. They are able to marshal their resources not only to survive, but also to prosper. A resilient structure or organzsation will be able to function in the face of adversity through a combination of good design, protection, effective emergency response, business continuity planning, and recovery arrangements. Examples of a lack of resilience include the 2010 Haiti earthquake when many buildings collapsed through poor quality construction, the explosions at Chernobyl in 1986 and the effects of the tsunami at the Fukushima nuclear plant in 2011.

At the heart of the assessment of risk and resilience are two difficult questions.

First, how do we know what we know? In other words how do we judge the 'pedigree' of information—particularly when we have to rely on it to make decisions about risks to the lives of others? How much can we rely on information from any source, be it scientific, heuristic (i.e. based on rules of thumb that have worked before), regulatory, commercial, or any other?

Second, in the 21st century we are beginning to realize an even more important, complex, and difficult question is—how can we know what we don't know? How could we have foreseen the attacks on the World Trade Center in New York before the event? How could we have foreseen the knock-on effects of the overtopping of the tsunami defences at Fukushima? How will we need to defend our infrastructure against a changing climate?

At the root of professional decision making, engineers have the same legal duty of care that all professionals have. Under the law of tort they have to take all reasonable care while performing any acts that could foreseeably harm others. It is the first thing to be established in any legal action against negligence. To carry out this duty engineers have to sort out 'information wheat' from 'information chaff'.

So for the first question—about the pedigree of information— engineers must have a clear idea, defensible in law, of what makes information dependable. Science is, of course, the major source. However, judgements have to be made when science is used in practical decision making because the *context* of the application may be very different from the context in which the theory was derived and tested. For example the behaviour of a column depends, as we have seen in Chapter 5, on end restraints. The idealized joints used to test theories in a laboratory are normally quite different from the bolted or welded end connections used in an actual structure.

You will recall that in Chapter 1 we introduced the limit states or limiting safe conditions of a structure and we divided them into serviceability and ultimate limit states. We said that the internal forces in all degrees of freedom should not exceed certain values of pulling or tension, pushing or compression, and sliding or shearing.

To show how this works Figure 21a–f illustrates the flow of forces through two different models of an aircraft fuselage—both somewhat idealized and simplified for illustration. Figure 21a is a pin-jointed truss and Figure 21c is a tubular or box beam, as discussed in Chapter 5 and Figure 18a–c. The central support is the landing undercarriage and the left hand support is a nose wheel. The downward applied loads represent the weight of the structure on the runway before take-off. Straightaway there are some difficult judgements. How heavy are the structure and its passengers and their luggage? How heavy is the fuselage and wing? How are these weights carried by the structure? Figure 21a shows a typical assumption that the total weight is transmitted to the truss at each of the bottom joints. The heaviest on the left includes the wings and fuel, and is forward of the landing wheels. The lighter loads are along the rest of the fuselage towards the tail.

If the joints in Figure 21a are pinned (i.e. the members are free to rotate) then each one has two degrees of freedom. However, in reality the joints will have some small unknown level of stiffness and this may create a small amount of bending. If we assume that the nose wheel is restrained horizontally then there are three restrained degrees of freedom at the support reactions and the pin-jointed truss is statically determinate. You will recall that in Chapter 4 we showed how, in this case, the triangle of forces can be used to calculate all of the forces at each joint in turn. In the diagram only the directions are shown. In Figure 21b we take the idea further forward by imagining a cut through the truss to create two slices—one to the left and one to the right. Because we have cut the truss the internal tension and compression forces are now effectively external forces. They are equal and opposite as

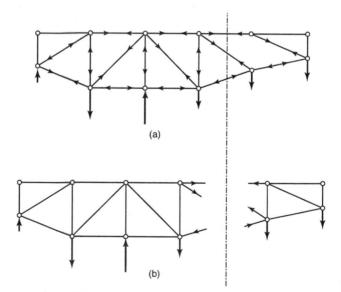

(a)

(b)

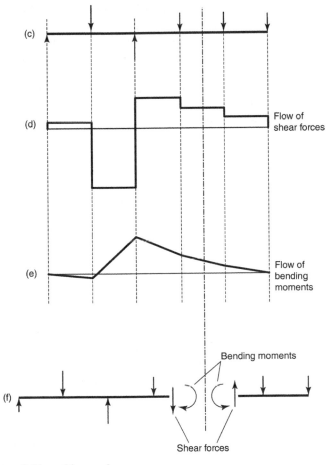

(c)

(d) Flow of shear forces

(e) Flow of bending moments

Bending moments

(f)

Shear forces

21c–f. Flow of forces—beam

they act on each slice and they must be in equilibrium with the other external forces applied to each slice separately.

We can calculate the vertical and horizontal forces equivalent to these internal forces at the cut (but not shown in the diagram).

Looking at the left hand slice, there will be vertical and horizontal forces at the top joint as well as at the bottom joint. The two vertical forces act along the same vertical line and when added together make up the total vertical shear force at the cut. The two horizontal forces, one at the top joint (pulling away from the slice) and one at the bottom joint (pushing in on the slice), are in opposite directions. Together they create a turning force, couple, or bending moment at the cut.

We can see how these same forces appear in the beam of Figure 21c. In Figure 21f we have imagined a cut in the beam at the same place to expose the shear forces and bending moments. We now think of these internal forces as external forces at the cut, and the vertical forces and their turning effects must balance. The diagrams in Figures 21d–e have been obtained by imagining cuts at different positions along the length of the beam. They show how the magnitudes of these internal shear forces and bending moments vary or flow along the length of the beam.

Several groups of limit states

So far we have only considered downward loading due to static weight on the runway. In other loading cases the applied forces will be acting in different directions. For example during flight the downward weight would be balanced or exceeded by a set of upwards lift forces, as shown in Figure 18a. Consequently over a range of very different situations the internal axial loads in each member in Figure 21a may at one time be tensile and at another be compressive. The structural engineers have to ensure that a member has the strength to take the maximum tensile and compressive internal loads over all of these possible circumstances or load cases. Such an analysis constitutes a first group of limit states—direct forces at both the serviceability and ultimate limits. In each load case, the limit strength in tension will be a maximum stress and in compression will be a maximum buckling load, as we discussed in Chapter 5. For the beam of Figure 21c, the limit strengths will be maximum shears and bending moments that the structure can resist—again in the serviceability and ultimate states.

Structures have to be sufficiently stiff—we don't want aeroplanes, ships, and bridges to sag in the middle too much or skyscrapers to sway too far. These are a second group of limit states. The deflections of the structure at each of the joints in each of the load cases in Figure 21a–f have to be calculated and checked. A third group of limit states is related directly to stiffness and requires that the structure does not vibrate or oscillate in a manner that causes discomfort to people or threaten the safety of the structure. In Chapter 5 we saw that the dynamic behaviour of an aeroplane, ship, or large bridge, such as resonance and flutter, can be quite complex. The engineers have to ensure that the structure is able to withstand forces due to movement such as when an aeroplane takes off or lands, or when a bridge is blown by the wind. They must ensure that the natural frequency of the structure is an acceptable value. Other complex behaviours have to be controlled—such as vortex shedding where eddies or vortices around a bluff body cause it to oscillate, buffeting due to shock waves, galloping at low frequencies, and other sudden impact loads, such as trucks crashing into bridge supports.

As we discovered in Chapter 5, varying loads such as traffic travelling over a bridge, wind gusting on an aeroplane, or wave action on a ship can cause fatigue damage. Fatigue is one of a fourth group of limit states which is concerned with deterioration over time. Other limit states in this group are degrees of corrosion and cracking from a variety of causes including pollution, electrochemical effects such as rust, damage by frost, insects, fungi, and bacteria. A fifth group of limit states includes discrete events, for example from the weather (lightning, heavy rain, snow, and hail stones, etc.); for aeroplanes, a tyre burst or a bird strike; and, for a ship, running aground. A sixth group of limit states specifically for aircraft, ships, and vehicles are called crashworthiness. These are limiting conditions that directly protect people during impact through seat belts, strong floors, fuel containment, and ensuring, as far as possible, the integrity of the structure if, for example, a plane is ditched.

Experience from previous similar jobs is invaluable to help deal with these complex issues. Many heuristics or 'rules of thumb' are used in practice—sometimes even in rather simple cases. For example engineers often have to decide how close a hole can be drilled to the edge of a notch in a steel beam. They would probably use the simple rule that the distance should be not less than 1.5 times the diameter of the hole. They would not carry out a full scale scientific finite element analysis (Chapter 5) of the stresses. There are at least two reasons. First, it would seem unreasonably wasteful of resources—using a 'sledge hammer to crack a nut'. Second and more importantly, the answers would not be helpful. The engineer analyst would choose the sizes of the elements and the computer would produce some answers. Then in order to get a more accurate answer the engineer might make the elements smaller and carry out a new analysis. The computer would predict higher stresses. This process could be repeated. Each time the computer would predict higher stresses. In other words as the sizes of the finite elements are reduced, the computer predicts higher and higher stresses. The reason for this strange behaviour is that the theoretical stresses at a notch approach infinity (whereas the actual stresses become non-linear and plastic, as shown in Figure 16c). So the theoretical answer depends on the method—represented here by the size of the element. The lesson is that finite element analysis can be seriously misleading if not performed by someone who understands the approximations made in setting up the calculational model, and who can interpret the results from the computer. The way the modelling is done is critical for the dependable use of a powerful theoretical technique. In the wrong hands, theory can be at least misleading and at worst dangerous.

International and national government and state regulations, codes of practice, and official documents also play a crucial role.

They are instrumental in protecting the safety of you and me: the public. They also help engineers deal with some of the most difficult technical issues. Where specific guidance is given, then it defines the standard of the duty of care that engineers must reach. One example is the level of a factor of safety (the ratio of the strength to load) in a particular limit state. The factor of safety of a component in a building acting only under self-weight in a serviceability limit state might typically be 1.0, but 1.3 in the ultimate limit state. Equivalent figures for moving traffic loads on a bridge might be 1.2 and 1.7. For aircraft, they may be 1.0 and 1.5.

Just one example, for ships, will illustrate some of the detail in these rules and regulations. The Lloyds Register Rules and Regulation for the Classification of Ships July 2010 is 2,779 pages long. It contains a large number of requirements, for example for the manufacture, testing, and certification of materials used in the design and construction of ship structures, machinery, and electrical systems. The document also gives rules for the design process—so, for example, Part 3 Chapter 4 Section 5 has a paragraph of Rule 5.3.2 which reads, 'Still water bending moments are to be calculated along the ship length. For these calculations, downward loads are to be taken as positive values and are to be integrated in the forward direction from the aft end of L. Hogging bending moments are positive.' Aircraft structures have to comply with the requirements, called CS (Certification Specification) documents, of airworthiness authorities such as the European Aviation Safety Agency. The Federal Aviation Administration in the USA issue similar documents called FAR (Federal Aviation Regulations). Civil structures must be built in compliance with Euro codes in Europe, American Standards in the USA, and national building regulations. In the UK, the Highways Agency requires roads and bridges to be designed in accordance with the Design Manual for Roads and Bridges (DMRB).

All calculations should be checked, but requirements and common practice do vary quite widely across countries and industries.

Independent checking systems are adopted in Germany, Japan, and California. In France, a Bureau de Contrôle is an independent firm licensed by the state to check calculations for certain classes of building structures. In the UK, the DMRB defines four categories of structure. Minor structures in categories 0 and 1 require an independent check by another engineer who may be from the same design team. Structures in category 2 can be checked by engineers who may be from a different part of the same organization. Major category 3 structures must be checked by a separate organization.

Most regulations also allow engineers to justify decisions that go beyond the guidance through further calculation and testing. For example almost all aircraft components are subject to rigorous testing in order to keep self-weight to a minimum and yet comply safely with the regulations.

As we saw in the last chapter the sheer size of some structures creates its own problems—for example the testing of prototypes of large structures is difficult and expensive. Few people would expect to build a large bridge or a tall skyscraper, and then test it to destruction before rebuilding it in its final form. Structures like these are one-offs. Some warships, cargo ships, and submarines are designed as a 'class' so that more than one can be built. Aircraft manufacturers do make and sell, typically, a few hundred or more aeroplanes based on one type, with variations for their airline customers. Weight is so important in aircraft production that it makes sense to spend quite large sums of money to make the structure as efficient as possible. The net result is that aircraft are expensive, requiring much more research and testing, and longer lead times than one-off structures. However, in every case, the business of procurement is complex and requires that the various parties involved, the client, designers, contractors, and users, cooperate to the maximum extent possible.

One way of delivering a degree of resilience is to make a structure fail-safe—to mitigate failure if it happens. A household electrical fuse is an everyday example. The fuse does not prevent failure, but it does prevent extreme consequences such as an electrical fire. Damage-tolerance is a similar concept. Damage is any physical harm that reduces the value of something. A damage-tolerant structure is one in which any damage can be accommodated at least for a short time until it can be dealt with. For example an aeroplane maintenance programme should detect and repair accidental damage, corrosion, and fatigue cracking before it causes serious in-flight failure. Whenever something goes wrong then lessons must be learned and applied because failure can result in large-scale loss of life.

We touched on another aspect of resilience in Chapter 4. We saw that a statically determinate structure has only one pathway along which its internal forces can flow. It is like a chain of links—fail one of the links and the whole chain breaks. Aircraft designers call it the safe-life design principle. Now it is only allowed for landing gear. The reason is that a chain cannot easily be made fail-safe or damage-tolerant unless all of the load paths can be inspected at sufficiently regular intervals. A statically indeterminate structure has more resilience because it has more than one force pathway—it is like a set of chains in parallel. If one link fails then we lose one chain but the other chains can take over—provided they are strong enough. So if one pathway is blocked then the forces can flow along another path. Again it is desirable, but not always totally feasible, that all of the load paths can be inspected at regular intervals.

Low chance, high consequence risks are particularly difficult to handle, and so here resilience becomes crucial. For example before the collapse of the World Trade Center in New York (9/11) no-one seriously contemplated the possibility of a jumbo jet being flown deliberately into a skyscraper. Aeroplane impact was considered, but not as a deliberate act of violence. The structure cascaded as it fell in a manner that surprised many structural engineers at the time. It prompted engineers all over the world to reconsider the phenomenon of progressive collapse. This happens if some damage, large or small, 'unzips' a structure. If the initial damage is small but causes a cascading progressive failure through the structure (rather like a line of dominoes falling onto each other) then the structure is particularly vulnerable and not robust.

To make a structure more resilient to low chance, high consequence risks, a structural engineer must identify all possibilities where some small amount of damage can cause disproportionate consequences. One of the most famous examples of this was the collapse of a block of apartments at Ronan Point in London in 1967. A relatively small gas explosion in a domestic cooker caused the whole side of a tower block to collapse. Subsequently the UK building regulations and codes of practice were radically amended. Small practical details that may seem trivial may be of immense importance in this way. There are many examples where a tiny failure has escalated into total disaster. Perhaps the most famous is the Space Shuttle *Challenger* which exploded minutes after take-off in 1986 due to the failure of an O-ring seal. All of this checking is hard enough when the risks are foreseeable but the difficulty is compounded many times over if the damage is difficult to foresee.

Partly as a consequence, in the 21st century we have begun to realize the importance of the second question first set by Plato—How can we know what we don't know? The best science can offer us is evidence, some agreed and some disputed. For example we have evidence that climate change will lead to more extreme weather events—but what will the future actually hold? Collectively we have begun to understand that complex behaviour can emerge from interactions between many simpler but highly interconnected processes. We are beginning to recognize new high impact risks through interdependencies we may not fully understand.

Of course engineers want to get the best out of a system—they want to optimize, and therein is the challenge. Unfortunately a solution that is optimal or highly tuned in one context may well be vulnerable in another.

Risk is in the future

Risk is in the future—it is the chance of an event that may cause harm and the consequences that follow. There is no such thing as zero risk. No matter how remote a risk might be, it could just turn up. Risks have to be reduced or managed to a tolerable level. To tolerate a risk means that we do not regard it as negligible or something we might ignore, but rather as something we need to keep under review and reduce still further if and as we can.

Although engineers work with known scientific knowledge, we have seen that there are still many assumptions that have to be made about the nature of the limit states and how they may be exceeded. One way of categorizing the uncertainties is into three types, parameter, system, and human. Parameter uncertainties concern our estimates of quantities like the weight of a passenger and his or her luggage. We know that if we measure the weight of something many times over then we don't always get the same answer each time—mathematicians call it random error—a lack of repeatability of a measurement. More generally, randomness is a lack of specific order in some information. It describes variations that we cannot explain any other way—they just are. The tosses of a coin come out roughly 50:50 for each side of the coin over a large number of throws, but regarding any single throw we cannot be sure which we will get—heads or tails. So randomness is all around us. The weights of an aeroplane, a ship, a bridge, and all the components in all structures can't be known exactly. Other quantities such as the sizes of members and their material properties (e.g. their yield stresses and their stiffness coefficients) vary. Random variations in external loads are even greater. They can be captured using statistical techniques—for example, the famous 'bell shape' of the normal distribution used in probability theory. Some structural engineering scientists have developed theories of structural reliability which manipulate these statistics to predict probabilities of failure—in other words the chance that a given applied load force overcomes the strength force in a given limit state. Unfortunately these calculations give scant attention to two other major sources of uncertainty—system and human—and so can be wildly misleading.

We have already quoted many examples of system uncertainty—they are to do with how we represent the physical world in our theoretical scientific models and interactions between these models at the different levels of the structural hierarchy. For example in Figure 21a we assumed the joints were pinned and that the weights were applied at the bottom nodes. We saw how the application of the finite element technique requires great skill, experience, and understanding if it is not to give misleading answers. We have seen the difficulties dealt with by the use of rules of thumb and regulations—they are all examples of system uncertainty.

The final category is the uncertainty due to human behaviour, both individually and collectively. Some man-made disasters have been due to human error such as calculational mistakes or errors of judgement, for example the sequence of decisions made before the Chernobyl nuclear accident. However, the more difficult human uncertainties derive from social interactions that have not been foreseen. Just as there are technical hazards so there are human and social hazards. Social scientists, such as Barry Turner, Nick Pidgeon, Charles Perrow, and Jim Reason, have studied many man-made disasters and failures, including structural failures such as the box girder bridge collapses in the 1970s. As we hinted earlier, they have discovered that human factors in failure are not just a matter of individuals' slips, lapses, or mistakes but are also the result of organizational and cultural situations which are not easy to identify in advance or even at the time. Indeed, they may only become apparent in hindsight.

It follows that another major part of safety is to design a structure so that it can be inspected, repaired, and maintained. Indeed all of the processes of creating a structure, whether conceiving, designing, making, or monitoring performance, have to be designed with sufficient resilience to accommodate unexpected events. In other words, safety is not something a system *has* (a property), rather it is something a system *does* (a performance). Providing resilience is a form of control—a way of managing uncertainties and risks. It is a process about which past experience can give us some confidence that we can cover the uncertainties with suitable factors of safety and working practices. But we must always be aware of significant changes in context and the possibility of the unexpected.

As a consequence some engineers have begun to think differently, using what many call 'systems thinking' to deal with complex unintended consequences from unexpected interactions. By this view, a system is a combination of parts that form a whole in a hierarchy of levels of definition, as set out in Figure 2. In other words, the parts are wholes and the wholes are also parts—Arthur Koestler called them 'holons'.

Traditionally we separate as being quite different our understanding of 'hard systems' of physical material objects, such as a ship, from 'soft systems', which involve people—the passengers and crew. Hard systems are studied through the physical sciences and soft systems through the social sciences and management. Systems thinkers want to integrate them to better understand properties that may emerge from interactions between sub-processes. They want to identify situations where the whole is more than the sum of its parts, both to deal with emerging hazards and to take new opportunities to create synergy.

The hard system of the ship has a life cycle as it is conceived, designed, built, used, and eventually decommissioned. As we have seen in earlier chapters the ship's structure has external work done on it and responds by doing internal work as it performs. The physical ship is in effect a manipulator of energy through processes of change. Energy is stored, changed, and dissipated in ways that depend on how the parts of the system are connected—its structure.

Every hard system is embedded in a soft system—ultimately all of us. The physical world 'out there', outside all of us, exists, but we can only perceive it through our human senses. We live and act in the world by creating our own mental models. We share and discuss and test them through processes that enable us to come to some agreement and that help us to act in ways that seem appropriate. That is how the physical sciences and engineering develop. The ship's crew is a soft system that has the job of understanding the hard system of the ship, predicting how it may behave in various situations, and making sure as far as is possible that it can be controlled and managed safely throughout a voyage.

So by this thinking, resilience is an outcome of a process that *emerges* from the interactions between joined-up and integrated hard and soft sub-processes. Just as strain energy is an outcome, expressed as a property of a material that emerges from the interactions between its molecules, so the resilience of a project emerges from the interactions between its well engineered hard and soft sub-processes. A central aspect of vulnerability, and hence robustness, resilience, and sustainability, of technical and socio-technical systems, is how we ensure that emerging 'surprises' are managed, especially those that have high impact but are of low chance or probability. Progress to success is tracked explicitly by collecting and interpreting uncertain evidence to manage emerging risks. Attention is required on every process on every level in the hierarchy, in other words engineers have to maintain a strategic view whilst at the same time paying deep attention to the smallest detail.

We can draw a direct analogy between medical staff monitoring the symptoms of ill health or disease in people and engineers monitoring the symptoms of poor performance or proneness to failure in an engineering structure. Structural health monitoring is the taking of measurements of performance to detect changes that might indicate damage and potential harm before it becomes obvious and dangerous.

At the start of this chapter, we looked at how structural engineers face some big and difficult questions. 'What are the risks? How big are they? How safe is safe enough? A central message of this book is that engineers work with *known* scientific knowledge, but that all is not predetermined. Scientific knowledge is necessary but not sufficient—and therein lies the art and professional expertise of the engineer. Many assessments and assumptions have to be made about what may happen, because there is always some risk in the future. Whenever something is done, some act performed, some project started, there is risk. There is no such thing as zero risk—no matter how remote, it could materialize. Risks either have to be physically removed altogether or reduced or managed to a tolerable level.

We have seen that we can predict and hence control risks from some random uncertainties. But by their very nature the answers are partial because of system and human uncertainties. Predicting everything that might happen is a challenge—the challenge of foreseeing a complete life cycle. Various forms of deterioration such as corrosion and fatigue are often difficult to inspect and repair.

Systems thinking is beginning to provide a common language for hard and soft systems and the most promising way of re-integrating those professions that are fragmented by specialisms. However, there is still some way to go since the majority of engineers, architects, and other related professionals have yet to embrace this way of thinking.

One conclusion is for sure—good structural engineering can save orders of magnitude in cost and protect the lives of both the living and the yet to be born.

Glossary

Action	A demand on a structure such as a load. Also used to describe a force caused by a load.
Bending moment	The internal turning force in a beam generated when two ends of an element of material are rotated in opposite directions.
Buckling	A failure mode of a structure that usually occurs suddenly in compression where a structure bifurcates (i.e. divides into one of two states) and takes the path of least resistance by moving out of the line of action of the load.
Catenary	The curve formed by a flexible cable as it hangs under its own weight from two points.
Compatibility	The requirement that deformations and deflections of the parts of a structure are consistent with each other, e.g. member distortions must be consistent with the deflections.
Constitutive equations	Equations or rules expressing the way in which a material will behave.
Damping	The process by which a structure loses energy due to friction and a general 'looseness' between its parts.

Degrees of freedom	The number of independent coordinates that are needed to describe the configuration of a structure.
Elastic	The ability of a material to regain its original shape after being changed—for example by being pulled or compressed. The change may be linear or non-linear.
Energy	A capacity for work—potential energy is due to position; kinetic energy is due to movement; and strain energy is potential energy in a deformed material. These are measured in joules.
Finite element method	A computer method of analysing the behaviour of a structure, material, or fluid. It relies on splitting the structure up into a large number of discrete 'bits' or elements. Equations relating the internal and external forces and displacements for all the elements are set up and solved by a computer in a way that would be difficult to do manually.
Hogging	Bending upwards to form a shape that is concave below—the opposite of sagging.
Limit state	A condition beyond which a structure is deemed to have failed to achieve its fitness for purpose.
Load	An action applied to a structure—forms include dead load (self-weight), live loads (non-permanent loads such as snow—also called superimposed or imposed loads), wind loads, earthquake loads, and impact loads (collisions and wave slamming etc).
Metal fatigue	The deterioration of the strength and other mechanical properties of a material when subjected to repeated applied stress in the elastic range. A normally ductile material may become brittle and crack.
Pinned joint	One in which the structural members can rotate freely with respect to each other.

A pinned support to a structure is one in which the members may rotate with respect to the support (though they may or may not be pinned to each other).

Reaction	An opposition to an action—an applied force.
Resilience	The ability of a system to respond to a demand (e.g. to detect, prevent and manage disruptive challenges).
Resonance	Where the frequency of a stimulus is close to the natural vibration frequency and causes very large vibrations.
Risk	The chance or likelihood of a specified situation, at some time in the future, set in a stated context, and combined with the consequences that will follow.
Sagging	Bending downward to form a shape that is concave above—the opposite of hogging.
Scalar	A quantity with magnitude but not direction; *see also* Vector.
Shear force	An action of sliding one layer of material over another.
Shell	A structure with the form of a thin, curved sheet or plate. A grid shell is a shell with regular holes like the wire mesh of a kitchen sieve.
Statically determinate	Where the internal forces in a structure can be found by balancing the forces (simple statics) alone.
Statically indeterminate	Where the internal forces in a structure cannot be found by balancing the forces (simple statics) alone.
Stiffness	The relationship between load and deformation of a structural member.
Strain	The deformation produced by a load.
Stress	The force on a given area of material.

Systems thinking	A way or philosophy of approach to problem-solving that values both wholes and parts (i.e. that combines reductionism with holism).
Vector	A quantity with direction as well as magnitude.
Virtual work	Virtual means not physically existing. Virtual work is work done in a thought experiment on a structure when a force is moved through an imaginary virtual displacement. That virtual displacement may or may not be the real displacement in the actual structure.
Vulnerability	Susceptibility to small damage causing disproportionate consequences.
Work	A transfer of energy as a force moves a distance (i.e. force times distance—measured in joules).

References

Chapter 1: Everything has structure

The microstructure of spider silk is being investigated for a number of potential uses including bullet-proof vests and artificial tendons.

For a discussion of the idea of systems architecting within systems engineering, see for example:
Michael R. Emes, Peter A. Bryant, Michael K. Wilkinson, Paul King, Ady M. James, Stuart Arnold, Interpreting Systems Architecting. *Systems Engineering* Vol. 15, No. 4, 2012, pp. 369–95. Available at <http://onlinelibrary.wiley.com/doi/10.1002/sys.21202/abstract>. Last accessed March 2014.

Vitruvius: *Ten Books of Architecture*, ed. Ingrid Rowland and Howe Thomas (Cambridge University Press, 2001).

Chapter 2: Does form follow function?

In everyday usage, mass and weight are interchangeable because gravity is more or less the same everywhere on the surface of the Earth (though it does vary a little). So we speak of our own body weight in kilograms (kg), for example, even though a kg is a unit of mass. Scientists make the distinction clear because on the Moon and planets the acceleration due to gravity is different. So, on the Moon, whilst your mass would be the same your weight would be different. Confusingly scientists sometimes express weight as a kilogram force (kgf) which is the weight of a mass of 1 kg = 1 kg *9.81 metres (m)/

second² or approx 10 Newtons (N). A weight of 10 kiloNewtons (kN) is 10,000 N and is the weight of a mass of 1,000 kg or 1,000 kgf. And 1,000 kg is one metric ton or tonne. So 10 kN is the weight of a 1 tonne mass or 1 tonne force. It is very approximately the weight of the mass of an English ton or 2,240 pounds (lbs).

Stress is a force over a defined area—for example 100 N per square millimetre (mm).

E. Happold, Timber Lattice Roof for the Mannheim Bundesgarten. *The Structural Engineer*, Vol. 53, No. 3, March 1975.

A bridge is embedded in and connected to its environmental, human, social, political, and cultural context.

Chapter 3: From Stonehenge to skyscrapers

A *henge* just means a ditch or earthwork—but such a simple description diminishes the incredibly strong beliefs that drove the enormous effort required. At first Stonehenge was just a simple circle, about 87 m diameter, of round pits each about 1 m wide and deep. Around 2150 BC, 82 so-called Bluestones, of up to 4 tonnes each, were moved approximately 250 miles from south-west Wales and set up in an incomplete double circle. About 2000 BC heavier Sarsen stones were brought from the Marlborough Downs about 25 miles away. Then in about 1500 BC the stones were rearranged in a horseshoe and circle. Over the years many stones have been removed or broken up, and some remain only as stumps below ground level.

An outline of the French theory about how the pyramids were built can be found at <http://news.nationalgeographic.com/news/2008/11/081114-pyramid-room.html>. Last accessed March 2014.

Mohr's circle is a circle on a two dimensional graph. The axes of the graph are normal and shear stresses. Each point on the circumference of the circle is the stress on a particular plane.

C. P. Snow, *The Two Cultures and the Scientific Revolution* (Cambridge University Press, 1959): first delivered as a lecture about the gap between the arts and the sciences.

Chapter 4: Understanding structure

A cross section is the shape of a body cut through transversely to its length. The cross section of a tower is the part that you see if you imagine a cut horizontally through the tower.

A centre of gravity is the point in a body at which the total weight may be considered to act.

We recognize that we can loosen a cane, which has been pushed into the ground, by wobbling it back and forth—a complication that is important in structures buffeted by wind and earthquakes.

In the mathematics of infinitesimal calculus δx is a standard way of expressing the idea that the element is of very small length.

The proof that you can always stabilize a table by turning it only works where the floor is uneven but has no steps and the legs are of equal length. If there is a step or the legs are unequal then you may be able to find a stable position, but we cannot prove you always will—you may well still need table napkin packing! The reasoning why turning works on a rough continuous floor is as follows. Our rough surface floor will have an average level that is flat. We know that particular points on the rough floor will sometimes be above that level and sometimes below it. We know we can always find a position with three legs grounded. So let's imagine the table has three legs of fixed equal length but the fourth one has a spring on its foot—so its length can vary. As we turn the table the four legs will be moving round a ring on the floor and the spring on the fourth leg will move in and out. Because it is sometimes longer and sometimes shorter than the others, there must necessarily be some point when it is exactly the same length. Therefore at that angle of rotation the table is touching the ground on all four legs of equal length.

Chapter 6: Resilience

As far as I can trace back the whimsical definition of structural engineering first appeared in E. H. Brown, *Structural Analysis*, Volume 1 (Longmans, 1967).

See Arthur Koestler, *The Ghost in the Machine* (Picador, 1975) for the idea of a holon.

Further reading

Chapter 1: Everything has structure

Nigel Hawkes, *Structures: Man-Made Wonders of the World* (Readers Digest, 1990). A coffee table book with good photographs and descriptions.

David Yeomans, *How Structures Work* (Wiley-Blackwell, 2009). An introduction based on equilibrium through simple statics—aimed at architects with a minimum of mathematics.

Spiro Kostof (ed.), *The Architect: Chapters in the History of the Profession* (University of California Press, 1977). Covers the development of architecture from ancient to modern.

Andrew Ballantyne, *Architecture: A Very Short Introduction* (Oxford University Press, 2002).

Chapter 2: Does form follow function?

Gijs Van Hensbergen, *Gaudí* (HarperCollins, 2002). A well written biography of Gaudí.

Wanda J. Lewis, *Tension Structures: Form and Behaviour* (Thomas Telford, 2003). A technical book.

William Addis, *The Art of the Structural Engineer* (Artemis, 1994). Demonstrates the creative role of the structural engineer.

Ken Baynes and Francis Pugh, *The Art of the Engineer* (Lutterworth Press, 1981). Celebrates the art of engineering drawings.

Chapter 3: From Stonehenge to skyscrapers

Kurt Mendelssohn, *The Riddle of the Pyramids* (Thames and Hudson, 1974). An Oxford physics professor looks at the pyramids.

Robert Benaim, Engineering Architecture, *Ingenia* 11 February 2002. Available at <http://www.ingenia.org.uk/ingenia/issues/issue11/benaim.pdf>. Last accessed March 2014. Argues that architecture and engineering have been artificially separated.

David Billington, *The Tower and the Bridge* (Princeton University Press, 1985). Celebrates the Eiffel Tower and the Brooklyn Bridge along with many other structures to identify the art of structural engineering.

Larrie D. Ferreiro, *Ships and Science* (MIT Press, 2010). Tells the story of the birth of naval architecture.

Chapter 4: Understanding structure

Structural Engineering

David Blockley, *Bridges, the Science and Art of the World's Most Inspiring Structures* (Oxford University Press, 2010). The science and art of bridges.

J. E. Gordon, *The Science of Strong Materials* (Pelican Books, 1968). J. E. Gordon, *Structures: or Why Things Don't Fall Down* (Penguin Books, 1978). These two books are best-selling classic texts—well written accounts for the general reader. However, they do not cover the principles required for computer based methods used for modern statically indeterminate structures nor do they cover risk, safety, and resilience adequately for the 21st-century reader.

Mario Salvadori, *Why Buildings Stand Up* (W. W. Norton & Co., 1980). Matthys Levy and Mario Salvadori, *Why Buildings Fall Down* (W. W. Norton & Co., 1994). As for Gordon's books, both of these books are well written accounts of some of the basics. However, they are potentially more misleading than helpful in that they include descriptions of advanced structures without sufficiently discriminating between statically determinate and indeterminate structures. Risk, safety, and resilience receive scant attention.

Chapter 5: Movers and shakers

Guy Norris and Mark Wagner, *Airbus 380: Superjumbo of the 21st Century* (Zenith Press, 2010). A coffee table book about the aircraft which describes the structure and how it was put together.

Ken J. Rawson and Eric C. Tupper, *Basic Ship Theory*, Volumes 1 and 2. 5th edn (Butterworth Heinemann, 2001). A simple text but with some mathematics.

John D. Anderson, Jr., *The Airplane: A History of its Technology* (American Institute of Aeronautics and Astronautics, 2002). Explains the basics in plain language—a good preparation for more advanced texts.

John Cutler, *Understanding Aircraft Structures*, 2nd edn (Blackwell Science, 1992). Explains how aircraft structures work—with some historical background.

Chapter 6: Resilience

David Blockley, *Engineering: A Very Short Introduction* (Oxford University Press, 2012). Chapter 6 contains a very brief introduction to systems thinking.

Barry A. Turner and Nick F. Pidgeon, *Man Made Disasters*, 2nd edn (Butterworth-Heinemann, 1998). Describes how human and organizational issues can incubate to cause accidents.

Peter Senge, *The Fifth Discipline* (Century Business, 1990). An excellent introduction to systems thinking.

David Blockley and Patrick Godfrey, *Doing it Differently* (Thomas Telford, 2000). Advocates systems thinking to deliver major projects.

Case studies of construction details and project management

O. J. Kelly, R. J. L. Harris, M. G. T. Dickson, and J. A. Rowe, Construction of the Downland Gridshell, *The Structural Engineer*, Vol. 79, No. 17, 4 September 2001, pp. 25–33. Available at <http://www.istructe.org/journal/volumes/volume-79-(published-in-2001)/issues/issue-17/articles/construction-of-the-downland-gridshell>. Last accessed March 2014. A good clear account of how it was all done.

Chris Wise, Andrew Weir, George Oats, and Peter Winslow, An amphitheatre for cycling: the design, analysis, and construction of the London 2012 Velodrome, *The Structural Engineer*, Vol. 90, No. 6, June 2012, pp. 13–25. Available at <http://www.istructe.org/journal/volumes/volume-90/issues/issue-6/articles/an-amphitheatre-for-cycling-the-design,-analysis-a>. Last accessed March 2014. An inspiring account of how this structural award winning structure was created and built.

Henry Petroski, *Design Paradigms: Case Histories of Error and Judgement in Engineering* (Cambridge University Press, 1994). Sets out how we have learned from some famous failures such as the Dee Bridge in 1848 and the Tacoma Narrows Bridge in 1940.

David Doran (ed.), *Eminent Civil Engineers* (Whittles Publishing Services, 1999). This book contains chapters written by engineers who are well known in their profession but not so well known outside of it. Each tells a personal story about what makes them tick, how they do their job, and why they find engineering so fascinating.

Alan Holgate, *The Art of Structural Engineering: The Work of Jörg Schlaich and his Team* (Edition Axel Menges, 1997). This is a big book of coffee table quality about Jörg Schlaich's life and work. Contains photographs and diagrams of construction details that show how his structures were put together. Examples include Jeddah Airport Haj Terminal Roof of 1982 and the Munich Olympic Cable Net Roof of 1972.

Will Jennings, *Olympic Risks* (Palgrave Macmillan, 2012). Traces how managing risk has become central to a successful games season.

The author's website, called Bridges, at <http://www.bristol.ac.uk/civilengineering/bridges/index.html> describes how bridges work as structures, and gives construction details for a number of well known and not so well known bridges around the world. Last accessed March 2014.

Index

SOCIAL MEDIA
Very Short Introduction

Join our community
www.oup.com/vsi

- Join us online at the official Very Short Introductions Facebook page.
- Access the thoughts and musings of our authors with our online blog.
- Sign up for our monthly e-newsletter to receive information on all new titles publishing that month.
- Browse the full range of Very Short Introductions online.
- Read extracts from the Introductions for free.
- Visit our library of Reading Guides. These guides, written by our expert authors will help you to question again, why you think what you think.
- If you are a teacher or lecturer you can order inspection copies quickly and simply via our website.